Erratum

Erratum to: Chapters 15
Tribology in Total Hip Arthroplasty

Karl Knahr (Ed.)

Erratum to: Chapter 15. Osteolysis and Aseptic Loosening: Cellular Events Near the Implant
DOI: 10.1007/978-3-642-19429-0_15

In the caption of figure 15.2 it should read 150 μm and not 150 mm.

The online versions of the original chapter can be found under
DOI: 10.1007/978-3-642-19429-0_15

K. Knahr (ed.), *Tribology in Total Hip Arthroplasty*,
DOI: 10.1007/978-3-642-19429-19, © 2011 EFORT

Tribology in Total Hip Arthroplasty

Karl Knahr

Editor

Tribology in
Total Hip Arthroplasty

Editor
Prim. Univ. Prof. Dr. Karl Knahr
Orthopaedic Hospital Vienna-Speising
2nd Orthopaedic Department
Speisinger Straße 109
1130 Vienna
Austria
karl.knahr@oss.at

ISBN 978-3-642-19428-3   e-ISBN 978-3-642-19429-0
DOI 10.1007/978-3-642-19429-0
Springer Heidelberg Dordrecht London New York

Library of Congress Control Number: 2011930549

Cover design: eStudioCalamar, Figueres/Berlin

Printed on acid-free paper

Springer is part of Springer Science+Business Media (www.springer.com)

Preface

Successful long-term results of total hip arthroplasty are mainly due to two facts: long-term stability of the implant and minimal wear of the articulating surfaces. Nowadays fixation of the implant component appears to be a minor problem as both options – cementless fixation and fixation using fourth generation cementing techniques – can achieve excellent long-term stability.

Wear of the articulating components in total hip arthroplasty remains the most challenging unsolved problem. The ideal bearing surface for total hip arthroplasty has been sought since the early days of this procedure. Beginning with polyethylene as a bearing surface by Sir John Charnley, a metal-on-metal bearing surface was introduced by McKee and Farrar to improve wear characteristics aiming at long-term survival of the implant. In the early 1970s Boutin initiated the first ceramic-on-ceramic articulation in France. Since these pioneering activities improvements in manufacturing techniques and materials have led to better long-term results – nevertheless each of these bearings have not only strengths but also weaknesses.

Conventional polyethylene–metal articulations are complicated long-term by wear debris and subsequent osteolysis and loosening. During recent years cross-linked polyethylene has shown improved wear characteristics compared with conventional polyethylene. Recent reports suggest that we have not yet reached the final stage of improvement as today vitamin E–stabilised cross-linked polyethylene is increasingly launched on the market. But, wear probably will not be as low as hard-on-hard bearings and we do not know if these new poylethylenes really can eliminate osteolysis especially in young and active patients.

The major advantage of ceramic-on-ceramic is its very low wear. But – there is still concern because of squeaking and fracture of the ceramic components. During the last decades ceramic technology has improved dramatically. Starting from a fracture risk of 1% in the 1980's there is now a probability of fracture for the ceramic head of only 0.002% and for the ceramic inlay of 0.02%. This improvement in technology is still in progress. Current problems concerning ceramic-on-ceramic prostheses include squeaking phenomena leading to patient complaints and, in some cases, to revision of the articulation. Evaluation of all these cases could show that this problem is associated with special types of implants and imperfect surgical techniques, for example, stripe wear due to edge loading of the ceramic inlay.

Metal-on-metal prostheses show low wear, no fracture risk and allow the largest femoral head-to-outside-cup-diameter-ratio. However, the concern here is the systemic metal ion level elevation and metal allergy resulting in local lymphocytic response. Recent reports of increasing failure rates using this material in resurfacing arthroplasty as well as in large-diameter head metal-on-metal articulations have caused official warnings from some State authorities. Nevertheless, there are still high numbers of cases with excellent clinical results and no problems related to the metal articulation.

Considering all these advantages and disadvantages we have to be aware that wear issues are still a challenge in achieving the ultimate goal of total hip arthroplasty – an implant which functions for the whole life of every single patient.

Vienna, Austria   Prim. Univ. Prof. Dr. Karl Knahr

Contents

Part I

Basics in Tribology

Tribology of Hip Prostheses

1

John Fisher

1.1
Introduction

Tribology is the study of systems that move and, in particular, mechanics, friction, lubrication and wear. Wear of bearing surfaces and the resulting adverse reactions to wear products are the major causes of long-term failure in hip prostheses [7, 9, 14, 15]. In this chapter, our research studies of the tribology of hip prostheses over a period of 20 years are summarised, the wear of different types of bearings are compared under standard walking conditions and the wear of the same bearings under adverse conditions that can lead to increased wear and failure is described.

1.2
Polyethylene Acetabular Cups with Metallic or Ceramic Femoral Head Cups

Polyethylene acetabular cups coupled with metal or ceramic femoral heads remain the most popular bearing combination in the hip. In the long term, the polyethylene cups wear and the micron and submicron wear particles result in osteolysis and loosening [10, 14]. Particles accumulate in peri-prosthetic tissues until a critical volume and concentration is reached, which results in osteolysis. For example, a polyethylene wear rate of 30 mm^3/million cycles will reach a total wear volume of 500 mm^3 in between 8 and 16 years, depending on the level of activity, and this will result in osteolysis and failure in some patients. Patients have different levels of reactivity to polyethylene particles [21] and particle concentrations in tissues are dependent on access. It is important to reduce polyethylene wear rate in order to extend the osteolysis-free lifetime. The wear rate of polyethylene is dependent on the sliding distance and hence the size of the femoral head [7]. More recently, wear has been

J. Fisher
iMBE, Institute of Medical and Biological Engineering, LMBRU, NIHR Leeds
Musculoskeletal Research Unit, University of Leeds, Leeds LS29JT, UK
e-mail: j.fisher@leeds.ac.uk

K. Knahr (ed.), *Tribology in Total Hip Arthroplasty*,
DOI: 10.1007/978-3-642-19429-0_1, © 2011 EFORT

shown to be dependent on the area of the polyethylene being worn. Increases in the head size from 28 to 36 mm can double the wear rate. This remains a challenge, when larger diameter heads are preferred, in order to increase stability and range of motion [9].

Damage to metallic femoral heads has been shown to increase polyethylene wear [4, 5, 28], and the introduction of ceramic femoral heads has been shown to reduce polyethylene wear. Historically, polyethylene sterilised with gamma irradiation in air, oxidised and degraded in the body, resulting in an increase in wear and in the number of submicron wear particles [5, 8]. Stabilised polyethylene was introduced to reduce oxidative degradation and wear and was subsequently followed by intentional cross linking and stabilisation of polyethylene. Cross-linked polyethylene has shown a substantial 50–80% reduction in wear rates in the laboratory compared to historical polyethylene, with additional benefit when used with ceramic femoral heads [11, 12]. Cross-linked polyethylene produces smaller and more reactive particles. Wear rates of less than 10 mm^3/million cycles are predicted to give osteolysis-free lifetimes of over 20 years and this provides a good solution for patients over the age of 60. For younger, more active patients with greater life expectancies, it is beneficial to consider alternative bearings.

1.3
Ceramic on Ceramic Bearings

Alumina on alumina ceramic bearings provide the lowest friction and wear of all bearing couples, with up to 50 times less than polyethylene, under standard walking conditions [22]. They also have significant clinical history [24]. More recently, ceramic matrix composite bearings in the form of Biolox Delta have been introduced, with increased toughness and reduced risk of fracture and these have shown even lower wear rates [1, 27]. Ceramic on ceramic is currently the lowest wearing bearing and is available in sizes 28–40 mm. It is most commonly implanted as a modular insert and as such requires significant wall thickness, so a 36 mm head can only be used in acetabular cups greater than 54 mm in diameter. It remains a preferred solution for high-demand, young and active patients.

1.4
Metal on Metal Bearings

Metal on metal bearings have also been introduced to address the low wear needs of young and active patients and to provide greater design flexibility (for example, mono block cups and surface replacements). Under standard walking conditions, the wear of metal on metal hips is low and less than 1 mm^3/million cycles, producing small nanometre-size particles [3, 6, 30]. Substantial concerns remain about adverse reactions to metal wear particles particularly for high doses and when wear rates exceed 1 mm^3/million cycles. It has been shown that high metal ion levels or high particle concentrations can cause cell death and tissue necrosis [13, 15]. Friction of metal on metal bearings is higher than in other bearings [2]. Metal on metal bearings are lubricated in the mixed regime and wear is reduced by a protein boundary layer [17]. After the initial 'bedding-in' period, the bearing surfaces become more conforming, the contact stresses reduce and the protective protein boundary layer is formed [33] and the steady state wear of a metal bearing drops to well below 1 mm^3/million cycles [6, 30]. The diameter of the bearing influences these low wear conditions. A 36 mm bearing has a lower steady state wear than a 28 mm bearing, but for sizes above 36 mm there is little difference in steady state wear [18]. A larger diameter bearing has lower bedding-in wear due to higher conformity than smaller diameter bearing. A smaller clearance between head and cup decreases the initial bedding wear, but has little effect on steady state wear [9]. The influence of metallurgy has been debated extensively. It is preferable to use a high carbon alloy, with cast, wrought and heat-treated variants of high carbon alloys all producing low wear under standard conditions [25]. These low wear conditions with wear rates less than 1 mm^3/million cycles occur in over 85% of patients, resulting in individual clinical metal ion levels of less than 5 ppb.

1.5
Ceramic on Metal Bearings

Ceramic on metal bearings, a differential hard on hard bearing, was introduced recently to reduce wear compared to metal on metal bearings, to provide design flexibility on the acetabular side by using a metal insert, to allow 36 mm heads to be used in patients with size 50 mm acetabular sockets and to eliminate ceramic insert chipping. The wear and friction of ceramic on metal bearings has been shown to be less than metal on metal bearings and similar to ceramic on ceramic bearings [32]. The bearing has shown reduced metal ion levels compared to metal on metal bearings clinically [16, 32].

1.6
Wear Under Adverse Conditions with Head–Cup Rim Contact

The wear conditions described above, under the standard walking cycle, are generated when the centre of head and the centre of the cup are concentric, when the cup is correctly positioned with respect to biomechanical loading axis and the contact patch and the wear occurs within the articulating surface of the cup. Any deviation from these conditions can result in the tribological contact patch of the head contacting the rim (or edge) of the cup producing 'head–cup rim contact'. When the head contacts the rim of the cup, 'stripe wear of head and rim wear of the cup' occur. Under these adverse conditions wear may increase.

There are a number of different conditions that can produce 'head–cup rim contact'. Both the head and cup have six independent degrees of freedom, three rotational and three translational. Contact between the head and rim of cup can result from a number of different types of conditions. These include:

Translational mal-position

- Medial or superior translation of the centre of cup, failure to restore the cup centre, leading to translational joint laxity.
- Offset deficiency, failure to restore head centre, leading to joint laxity.
- Head–neck impingement and lever out of the head on to the rim of cup.

Rotational mal-position of the cup

- Mal-position of the cup (inclination) such that the rim of the cup intersects the tribological contact patch.

It is important to consider the impact of these conditions on the wear and wear mechanisms in the different types of bearing, as an increase in wear under adverse conditions may lead to an increased failure rate. The wear performance of each of the bearing combinations considered in Sect. 1.2–1.5 are now considered under these adverse conditions.

Metal on polyethylene bearings do not produce stripe wear on the head due to head–cup rim contact, as the rim of the polyethylene cup is softer than the head. The rim of the polyethylene cup is plastically deformed, but surface wear of the polyethylene is not increased [29]. The major concern of head–cup rim contact in polyethylene cups is fatigue damage and fracture due to high stress levels. This was evident in historical polyethylene that oxidised and remains a concern with cross-linked polyethylene today.

Ceramic on ceramic bearings show considerable resistance to head–cup rim contact. Rotational mal-position of the ceramic cup does not increase the very low wear rate [22]. Translational mal-position and microseparation produces stripe wear on the head and rim wear on the cup and a small increase in wear rate to approximately 1 mm^3/million cycles with alumina ceramic [23]. Ceramic matrix composite ceramic (Biolox Delta) shows even lower wear rates under microseparation conditions [1, 27]. The incidence of head–cup rim contact and stripe wear on femoral head is greater than 50% in retrieval studies [24].

Adverse conditions in the form of head–cup rim contact have emerged as a significant factor that cause increased wear [31], high metal ion levels and failure in up to 10% of metal on metal hips. Most (>85%) of the patients with metal on metal hips have low wear, less than 1 mm^3/million cycles and low metal ion levels less than 5 ppb. However, head–cup rim contact can increase wear by between 10 and 100 times. The amount of increase in wear and metal ion levels is dependent on the conditions and mechanisms that produce the head–cup rim contact and on component design. Different mechanisms of wear occur when the head contacts the cup rim. The contact stresses during head–cup rim contact are increased dramatically [20], the protective protein boundary layer is removed and there is evidence that mechanical (abrasive and adhesive) wear becomes more dominant. The wear becomes more aggressive and surfaces rougher. The damage to the head in the form of the stripe wear contacts the articulating surface of the cup and increases cup wear during normal articulation. As found with ceramic on ceramic bearings, there is evidence that head–cup rim wear produces larger micron-size particles [3], and that these particles may remain in peri-prosthetic tissues. Simulator studies show wear rates increased to between 1–5 mm³/million cycles with rotational mal-position and 1–10 mm³/million cycles with translational mal position or micro separation [19, 30–32]. This is a 10- to 100-fold increase in wear compared to low steady state wear rate and can lead to elevated clinical ion levels of between 5 and 50 ppb.

Under adverse head–rim contact conditions, ceramic on metal show a small increase in wear [32] similar to that found with ceramic on ceramic. The differential hardness means stripe wear on head is avoided, but there is a small increase in rim wear of the metal cup.

1.7
Discussion

Reduction in wear is a major factor in improving the long-term survivorship of hip prostheses, particularly in high-demand young and active patients. Bearings available today may well provide a lifetime solution for many patients. Cross-linked polyethylene on metal or ceramic femoral heads may provide a solution for most patients over the age of 60. These bearings are predicted to provide an osteolysis- free wear life of more than 20 years, and also are robust and tolerate clinical conditions that lead to head–cup rim contact. Alternative bearings, ceramic on ceramic, metal on metal and ceramic on metal, provide low wear under standard walking conditions. Ceramic on ceramic and ceramic on metal provide low wear under adverse conditions of head–cup rim contact. However, metal on metal bearings show substantially increased wear under adverse conditions of head–cup rim contact and this is a likely cause of increased failure rates in some designs of metal on metal bearings. All bearings need clinical follow-up to establish that the wear performance demonstrated in the laboratory is effective under different conditions found in different patients in vivo.

Acknowledgements   Research was supported by EPSRC, the Leeds Centre of Excellence in Medical Engineering, WELMEC, funded by the Welcome Trust and EPSRC, WT 088908/Z/09/Z and the Leeds Musculoskeletal Biomedical Research Unit (LMBRU), funded by NIHR. JF is an NIHR senior investigator.

References

1. Al-Hajjar, M., Leslie, I.J., Tipper, J., Jennings, L.M., Williams, S., Fisher, J.: Effect of cup inclination angle during microseparation on the wear of Biolox Delta ceramic on ceramic hip replacement. J. Biomed. Mater. Res. B Appl. Biomater. **95**, 263–268 (2010)
2. Brockett, C., Williams, S., Jin, Z.M., Isaac, G., Fisher, J.: Friction of total hip replacements with different bearings and loading conditions. J. Biomed. Mater. Res. B Appl. Biomater. **81**, 508–515 (2007)

3. Brown, C., Williams, S., Tipper, J.L., Fisher, J., Ingham, E.: Characterisation of wear particles produced by metal on metal and ceramic on metal hip prostheses under standard and microseparation simulation. J. Mater. Sci. Mater. Med. **18**, 819–827 (2007)

4. Barbour, P.S.M., Stone, M.H., Fisher, J.: A hip joint simulator study using new and physiologically scratched femoral heads with ultra-high molecular weight polyethylene acetabular cups. Proc. Inst. Mech. Eng. H **214**, 569–576 (2000)

5. Besong, A.A., Tipper, J.L., Ingham, E., Stone, M.H., Wroblewski, B.M., Fisher, J.: Quantitative comparison of wear debris from UHMWPE that has and has not been sterilised by gamma irradiation. J. Bone Joint Surg. Br. **80**, 340–344 (1998)

6. Firkins, P.J., Tipper, J.L., Saadatzadeh, M.R., Ingham, E., Stone, M.H., Farrar, R., Fisher, J.: Quantitative analysis of wear and wear debris from metal-on-metal hip prostheses tested in a physiological hip joint simulator. Biomed. Mater. Eng. **11**, 143–157 (2001)

7. Fisher, J.: Tribology of artificial joints. Proc. Inst. Mech. Eng. H **205**, 73–79 (1991)

8. Fisher, J., Chan, K.L., Hailey, J.L., Shaw, D., Stone, M.: Preliminary study of the effect of ageing following irradiation on the wear of ultrahigh-molecular-weight polyethylene. J. Arthroplasty **10**, 689–692 (1995)

9. Fisher, J., Jin, Z., Tipper, J., Stone, M., Ingham, E.: Tribology of alternative bearings. Clin. Orthop. Relat. Res. **453**, 25–34 (2006)

10. Green, T.R., Fisher, J., Stone, M.H., Wroblewski, B.M., Ingham, E.: Polyethylene particles of a 'critical size' are necessary for the induction of cytokines by macrophages *in vitro*. Biomaterials **19**, 2297–2302 (1998)

11. Galvin, A.L., Tipper, J.L., Jennings, L.M., Stone, M.H., Jin, Z.M., Ingham, E., Fisher, J.: Wear and biological activity of highly crosslinked polyethylene in the hip under low serum protein concentrations. Proc. Inst. Mech. Eng. H **221**, 1–10 (2007)

12. Galvin, A.L., Jennings, L.M., Tipper, J.L., Ingham, E., Fisher, J.: Wear of highly cross linked polyethylene against cobalt chrome and ceramic femoral heads. Proc. Inst. Mech. Eng. H **224**, 1175–1183 (2010)

13. Germain, M.A., Hatton, A., Williams, S., Matthews, J.B., Stone, M.H., Fisher, J., Ingham, E.: Comparison of the cytotoxicity of clinically relevant cobalt-chromium and alumina ceramic wear particles *in vitro*. Biomaterials **24**, 469–479 (2003)

14. Ingham, E., Fisher, J.: Biological reactions to wear debris in total joint replacement. Proc. Inst. Mech. Eng. H **214**, 21–37 (2000)

15. Ingham, E., Fisher, J.: The role of macrophages in osteolysis of total joint replacement. Biomaterials **26**, 1271–1286 (2005)

16. Isaac, G.H., Brockett, C., Breckon, A., van der Jagt, D., Williams, S., Hardaker, C., Fisher, J., Schepers, A.: Ceramic-on-metal bearings in total hip replacement: whole blood metal ion levels and analysis of retrieved components. J. Bone Joint Surg. Br. **91**, 1134–1141 (2009)

17. Jin, Z.M., Dowson, D., Fisher, J.: Analysis of fluid film lubrication in artificial hip joint replacements with surfaces of high elastic modulus. Proc. Inst. Mech. Eng. H **211**, 247–256 (1997)

18. Leslie, I., Williams, S., Brown, C., Isaac, G., Jin, Z., Ingham, E., Fisher, J.: Effect of bearing size on the long-term wear, wear debris, and ion levels of large diameter metal-on-metal hip replacements-an in vitro study. J. Biomed. Mater. Res. B Appl. Biomater. **87**, 163–172 (2008)

19. Leslie, I.J., Williams, S., Isaac, G., Ingham, E., Fisher, J.: High cup angle and micro-separation increase the wear of hip surface replacements. Clin. Orthop. Relat. Res. **467–469**, 2259–2265 (2009)

20. Mak, M.M., Besong, A.A., Jin, Z.M., Fisher, J.: Effect of microseparation on contact mechanics in ceramic-on-ceramic hip joint replacements. Proc. Inst. Mech. Eng. H **216**, 403–408 (2002)

21. Matthews, J.B., Green, T.R., Stone, M.H., Wroblewski, B.M., Fisher, J., Ingham, E.: Comparison of the response of primary human peripheral blood mononuclear phagocytes from different donors to challenge with model polyethylene particles of known size and dose. Biomaterials **21**, 2033–2044 (2000)

22. Nevelos, J.E., Ingham, E., Doyle, C., Nevelos, A.B., Fisher, J.: The influence of acetabular cup angle on the wear of "BIOLOX Forte" alumina ceramic bearing couples in a hip joint simulator. J. Mater. Sci. Mater. Med. **12**, 141–144 (2000)
23. Nevelos, J., Ingham, E., Doyle, C., Streicher, R., Nevelos, A., Walter, W., Fisher, J.: Microseparation of the centers of alumina-alumina artificial hip joints during simulator testing produces clinically relevant wear rates and patterns. J. Arthroplasty **15**, 793–795 (2000)
24. Nevelos, J.E., Prudhommeaux, F., Hamadouche, M., Doyle, C., Ingham, E., Meunier, A., Nevelos, A.B., Sedel, L., Fisher, J.: Comparative analysis of two different types of alumina-alumina hip prosthesis retrieved for aseptic loosening. J. Bone Joint Surg. Br. **83**, 598–603 (2001)
25. Nevelos, J., Shelton, J.C., Fisher, J.: Metallurgical considerations in the wear of metal-on-metal hip bearings. Hip Int. **14**, 1–10 (2004)
26. Stewart, T., Tipper, J.L., Streicher, R., Ingham, E., Fisher, J.: Long-term wear or HIPed alumina on alumina bearings for THR under microseparation conditions. J.. Mater. Sci. Mater. Med. **12**, 1053–1056 (2001)
27. Stewart, T.D., Tipper, J.L., Insley, G., Streicher, R.M., Ingham, E., Fisher, J.: Long-term wear of ceramic matrix composite materials for hip prostheses under severe swing phase microseparation. J. Biomed. Mater. Res. B Appl. Biomater. **66**, 567–573 (2003)
28. Tipper, J.L., Ingham, E., Hailey, J.L., Besong, A.A., Fisher, J., Wroblewski, B.M., Stone, M.H.: Quantitative analysis of polyethylene wear debris, wear rate and head damage in retrieved Charnley hip prostheses. J. Mater. Sci. Mater. Med. **11**, 117–124 (2000)
29. Williams, S., Butterfield, M., Stewart, T., Ingham, E., Stone, M.H., Fisher, J.: Wear and deformation of ceramic-on-polyethylene total hip replacements with joint laxity and swing phase microseparation. Proc. Inst. Mech. Eng. H **217**, 147–153 (2003)
30. Williams, S., Stewart, T.D., Ingham, E., Stone, M.H., Fisher, J.: Metal-on-metal bearing wear with different swing phase loads. J. Biomed. Mater. Res. B Appl. Biomater. **70**(2), 233–239 (2004)
31. Williams, S., Leslie, I., Isaac, G., Jin, Z., Ingham, E., Fisher, J.: Tribology and wear of metal-on-metal hip prostheses: influence of cup angle and head position. J. Bone Joint Surg. Am. **90**(Suppl 3), 111–117 (2008)
32. Williams, S., Schepers, A., Isaac, G., Hardaker, C., Ingham, E., van der Jagt, D., Breckon, A., Fisher, J.: Ceramic-on-metal hip arthroplasty a comparative in vitro and in vivo study. Clin. Orthop. Relat. Res. **465**, 23–32 (2007)
33. Yan, Y., Neville, A., Dowson, D., Williams, S., Fisher, J.: Tribo-corrosion analysis of wear and metal ion release interactions from metal-on-metal and ceramic-on-metal contacts for the application in artificial hip prostheses. Proc. Inst. Mech. Eng. J. **222**, 483–492 (2008)

Biomechanics of Hip Arthroplasty

2

Michael M. Morlock, Nick Bishop, and Gerd Huber

2.1
The Historical Perspective of Hip Biomechanics

The biomechanics of the hip joint has been of great interest to researchers and clinicians since the early days of anatomical studies. Julius Wolff addressed the relation between the inner architecture of the bone and the functional loading already in the nineteenth century [31] and Friedrich Pauwels built the foundation for a mechanical approach to understand joint loading 65 years later [24]. Both researchers, despite dealing with very different questions (bone remodeling vs. fracture mechanics), are good examples for the spread of the biomechanics field. It can be defined as the science concerned with the internal and external forces acting on the human body and the effects produced by these forces [13]. Pauwels elaborated on the influence of valgus (steep) and varus (flat) anatomical position of the femoral neck, demonstrating for a given joint position, that a valgus neck is associated with a smaller lever arm of the abductor muscles and larger abductor muscle forces. This increases the magnitude of the resultant hip joint force and also changes its point of action in the pelvis to a more lateral position. His findings influenced the treatment of femoral neck fractures and femoral osteotomies in a major way. The biomechanical situation is more complicated when applied to Total Hip Replacement (THA) since all joint parameters are influenced by the operation: joint center, neck angle, offset, lever arms, and the range of motion until impingement. Range of motion and joint stability are decisive issues, especially in younger patients with high expectations on their quality of life after THA. The varus and valgus situation as well as the hip joint center are determined by the position of the implant in the pelvis and femur. This positioning also influences the local loading situation at the implant component – bone interface. For example, a slightly superior, posterior, and medial hip joint center after replacement can be associated with markedly higher joint forces (Fig. 2.1).

M.M. Morlock (✉), N. Bishop, and G. Huber
Biomechanics Section, TUHH Hamburg University of Technology,
Denickestrasse 15, Hamburg 21073, Germany
e-mail: morlock@tuhh.de

K. Knahr (ed.), *Tribology in Total Hip Arthroplasty*,
DOI: 10.1007/978-3-642-19429-0_2, © 2011 EFORT

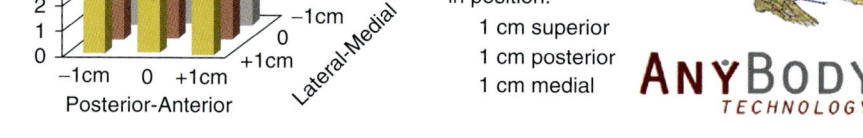

Influence of position of acetabular cup on hip reaction forces

Max. increase of hip reaction force: +60% in position:

1 cm superior
1 cm posterior
1 cm medial

AN**Y**BODY
TECHNOLOGY

Fig. 2.1 Simulation of the influence of acetabular cup position on the resultant hip joint reaction force (Courtesy of Sebastian Dendorfer, Anybody technology)

In vivo measurements with an instrumented femoral stem showed that dynamic hip joint forces lie in a similar range to those calculated by Pauwels with his static approach. However, for high body weight (BW) and an unfavorable loading situation (25% higher than average), peak forces during walking (3.9*BW), stair climbing (4.2*BW), and stumbling (11*BW) are substantial [5]. These high forces can be a factor in the failure scenario of hip joint replacements, especially if implant specific characteristics (e.g., small surface area) additionally increase the stress at the interface between bone and prosthesis components. The same applies to the junctions of modular prostheses systems. Aggravating in this context is also the continuously increasing body weight in most of the industrial countries. Not only the forces, but also the motion of the joint can play a role in failure scenarios of THA. The range of motion (RoM) at the hip utilized during normal daily activities is already quite substantial: flexion/extension can reach up to 124°, abduction/adduction up to 28°, and internal/external rotation up to 33° [14]. The athletic activities being performed by some patients with THA spans from running, cycling, kick boxing, alpine skiing to free climbing – activities, for which the RoM is most certainly higher.

In this chapter, an overview of the important factors influencing the function and longevity of THAs is attempted. For the identification of these factors, the analysis of the failure reasons for THA is helpful: The National Joint Replacement Registry of the Australian Orthopaedic Association lists "loosening/lysis of a prosthesis component" as the main (30%) and "dislocation" as the second most frequent indication (28%) for revision surgery [3]. Consequently, the emphasis is put on those biomechanical aspects related to design and implantation procedure, which directly or indirectly influence the occurrence of loosening, lysis, or dislocation. These aspects (in no particular order) are:

- Range of motion
- Impingement implant fixation
- Tissue damage during implantation and tissue tension after THA
- Component orientation (stem, cup)
- Bearing material

The aspect "bearing material" is not addressed in this chapter since it is extensively covered in another chapter in this book.

2.2
Range of Motion

Range of motion is influenced heavily by prosthesis design. The number of different designs of THA prostheses used is very large and can only be roughly estimated at 10,000–100,000. Femoral components vary with regard to material, length, diameter, shape, surface structure, surface coating, fixation, and stem modularity (Fig. 2.2). All femoral components have in common that the ball head (either modular or monobloc) articulates with the acetabular component. A wide variety of head diameters ranging from 22.25 mm to approximately 60 mm is available (Fig. 2.2). Acetabular components vary mostly with respect to the fixation mechanism and the bearing material, whereas the shape of the

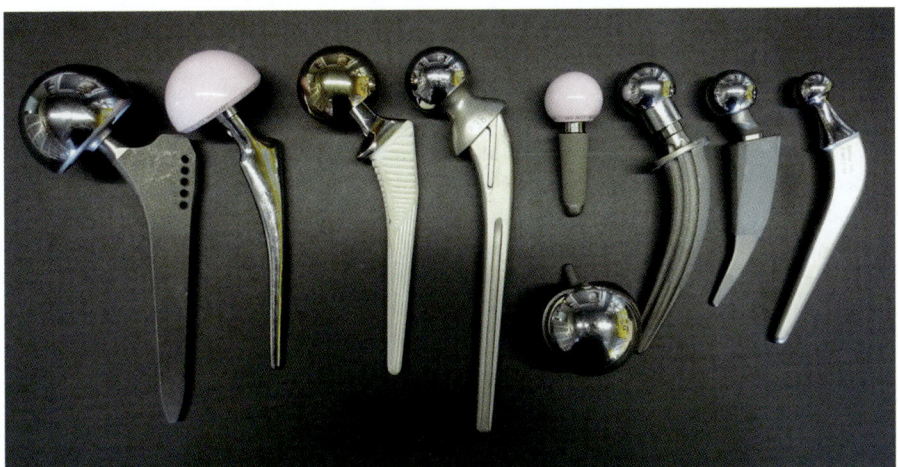

Fig. 2.2 Contemporary and historical femoral components documenting the wide variety of designs with head size diameters ranging from 62 to 22 mm (*left to right*: Zweymüller, Exeter, Corail, St. Georg, Silent & Resurfacing, CFP, Meta, Charnley)

Fig. 2.3 View of the backside of new and revised acetabular components; from left to right (*columns*): press-fit uncemented cups, cemented cups, threaded (*bottom*: expansion) cups, monobloc cups

acetabular cups is quite invariable, being hemispherical (sub-hemispherical) or conical in nature (Fig. 2.3). The possible head size is restricted by the outer diameter of the cup (i.e., the anatomical situation) and the required combined thickness of the cup shell and the bearing insert. In monobloc cups, shell and bearing insert are one piece and are, consequently, made from the same material (typically Polyethylene in the cemented and Cobalt-Chromium alloy in the uncemented case).

Head size directly influences the technical (theoretically possible) RoM. Increasing the head size from 28 mm to 36 mm yields an increase of 13° in the technical RoM (from 123° to 136°). This is derived for a hemispherical cup and a modern 12/14 mini taper completely embedded in the head (Fig. 2.4a) and a slender neck design (proximal diameter smaller than the distal diameter of the taper). The technical RoM is not directly related to the active or passive RoM achieved by the patient. This "true" RoM of the patient is heavily influenced by the orientation of the components, the muscular and soft tissue situation as well as the patient characteristics. Biomechanically most important is the position of the femur with respect to the pelvis, in which the end of the RoM is reached and the prosthesis neck "impinges" on the cup or impingement occurs somewhere else between femur and pelvis. Impingement can lead to sub-luxation or even dislocation of the hip joint. If impingement occurs repeatedly in positions inside the RoM required by the patient for either daily or athletic activities, dislocation is rather probable. In this situation, revision of the prosthesis (or at least one component of the prosthesis) is frequently required. The "jumping distance," which is the distance the head has to "jump" before leaving the cup, amounts in hemispherical cups to 50% of the head diameter. In sub-hemispherical cups, the distance is respectively less.

Fig. 2.4 Design aspects with direct influence on the range of motion [22]. (**a**) head size (*left: small, right: large*) (**b**) taper diameter (*large, small*) (**c**) cup entrance plane (*hemispherical, sub-hemispherical*)

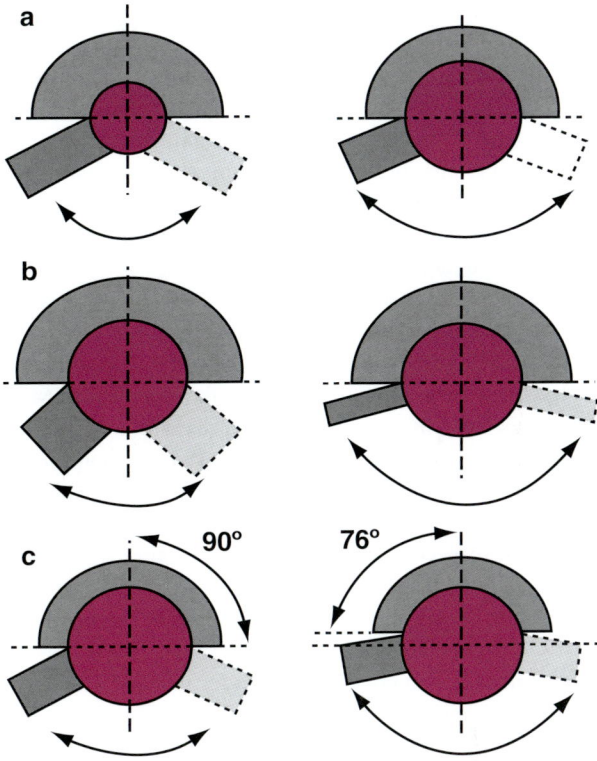

The advantage of larger heads with respect to the RoM and increased jumping distance is counter balanced by the higher friction moments, which have to be supported by the fixation of the bearing components. The friction increase with head diameter is pronounced for metal-on-metal articulations and worsened by the negative effect of resting periods on start-up friction with this material [8, 20]. The increased friction could be one of the important factors for the observed problems with cup loosening [17, 18] and taper corrosion [11]. A second disadvantage of larger heads is the higher separation of the joint that has to be achieved when relocating the head into the acetabulum. This separation corresponds to the jumping distance (in hemispherical cups). Consequently, the forces required to relocate the joint increase with larger heads. A weakening or damage of soft tissue structures can be the consequence.

Considering the advantages and disadvantages of large heads, the important question becomes: How large does it have to be? This question can only be answered by evaluating the clinical results. Nearly all publications document a decrease in the dislocation rate for an increase in head diameter. The absolute numbers, however, are quite different. For heads with a 28 mm diameter, they range from 0.6% [2] to 2.5% [25] or even 3.0% [6], 3.1% [1] and 3.6% [23]. For smaller head diameters the range is even wider: 3.8% [6] to 18.8% [23] for a 22 mm head. For larger head diameters, the rates are very low: for heads with 32 mm diameter only 0.5% [1], for 38 mm even 0.0% [25]. This indicates that the head diameter itself is only partly responsible for the dislocation rate. Implant position and soft tissue tension achieved by the surgeon are probably equally or even more important: "The theoretical gain in stability obtained by using a large femoral head (above 36 mm) is negligible in cases where there is a high cup abduction angle [27]."

Considering the pros and cons of large and extra-large heads, it is proposed that the head diameter is limited to about 36 mm for primary hip arthroplasty; in the case of Polyethylene possibly even to 32 mm, since for hard-soft bearings wear increases with head diameter. An initiative supporting and spreading this proposal was founded in 2008 by Carsten Perka from the Charité in Berlin and the first author of this paper and called "the 36 and under club."

The geometry of the stem taper is also an important aspect for the technical RoM: thinner tapers impinge later with the cup (Fig. 2.4b). The same applies to the neck geometry. Resurfacing of the femoral head is the extreme example for an extra-large diameter head with a thick neck. This explains why the technical RoM of resurfacing is about 31°–48° below the RoM of a stemmed prosthesis with a 32 mm head [16]. Available taper sizes range from 8/10 to 14/16 – the numbers correspond roughly to the proximal and distal diameter of the taper in mm. Thinner tapers have the disadvantage that the torque required to loosen the head on the taper decreases, which can cause disadvantageous rotation of the head with respect to the taper in high friction situations. It should be emphasized that tapers are not standardized, which makes the replacement of the head in a revision situation challenging. This problem is enhanced by the multitude of available sizes (8/10, 9/11, 10/12, V40, 11/13, C-taper, 12/14, 14/16).

A further design aspect important for the technical RoM is the location of the entrance plane of the cup and the cup profile [22]. In a typical hemispherical cup design with a hemispherical bearing liner, the center of rotation lies in the middle of the entrance plane and impingement occurs, when either the taper or the neck of the prosthesis come into contact with the implant (Fig. 2.4c). In implants with an elevated liner, the center of rotation lies below the cup entrance plane, as such reducing the RoM. The opposite effect is achieved by sub-hemispherical cup designs. In these designs, the cup only spans about 152°–166° instead of 180° [12]. In such designs, the center of rotation lies above (outside) the cup entrance plane [9]. The RoM is increased in sub-hemispherical cups, since impingement occurs later (Fig. 2.4c). The downside to this design variation is the decreased bearing surface, which is one of the factors made responsible for the increased wear in poorly functioning metal-on-metal prostheses.

2.3
Implant Fixation

THA implant components are fixed in the bone either by using cement or in an uncemented way with or without additional structures (e.g., screws). The choice of fixation method in THA varies greatly between countries: about 75% of all THAs in Sweden (2008), 51% in England & Wales (2003–2009) and 6.3% in Australia (2009), were implanted using cement for both components [3, 15, 21]. This indicates that the choice of fixation depends on many factors such as heritage of the surgeon, bone quality, and age among several others.

Table 2.1 Revision rates for the three commonly used fixation methods in THA

Revision rates by prosthesis type at one, three and five years for primary hip replacement procedures, undertaken between 1st April 2003 and 31st December 2009, which were linked to a HES/PEDW			
Prosthesis type	Number of patients	Revision rates (95% CI)		
		One year	Three years	Five years
Cemented	99,359	0.6% (0.6% to 0.7%)	1.4% (1.3% to 1.5%)	2.0% (1.8% to 2.1%)
Cementless	62,937	1.3% (1.2% to 1.4%)	2.5% (2.4% to 2.7%)	3.4% (3.2% to 3.7%)
Hybrid	31,662	0.9% (0.8% to 1.0%)	1.8% (1.6% to 1.9%)	2.7% (2.4% to 3.0%)

Adapted from [21]

The success of cementing relies heavily on the cementing technique, which has been continuously improved over the last two decades: vacuum-mixed cements, medullary canal plugs, centralizing elements, and the use of jet-lavage to clean the trabecular bone structures have been shown to effectively prolong the service life of prostheses [10]. Presently, cemented fixation still shows statistically the best results in terms of the whole THA population (Table 2.1). This changes when young and active patients are involved; in this patient collective, cemented prostheses do not perform as well as in the older population. This is the reason, why uncemented or hybrid fixation is common in this group.

The success of uncemented fixation depends on the ingrowth of bone. The ingrowth of bone is only possible, if the patient's activity-induced relative interface motion (micromotion) remains below a critical threshold in the early postoperative period (primary stability). Micromotion is induced by loading of the hip joint, which is almost impossible to avoid. In vivo measurements indicate that even static activities, such as lying in bed or working against resistance provided by the physiotherapist, create hip joint forces comparable to those occurring during unsupported walking [4]. The findings of Bergmann have challenged the advice of surgeons to patients to avoid full weight bearing and physical activities for the first few weeks after surgery. As a consequence, today only few surgeons still insist on partial weight bearing following total hip arthroplasty. It would appear that the quality of the initial fixation achieved by the surgeon, the characteristics of the implant surface, and the quality of the reamed or broached bony bed are more critical factors in achieving successful bone ingrowth than the influence of patient loads. The most frequently used method in uncemented implantations is "press-fitting." This method involves impaction of an implant into a cavity that is slightly smaller than the implant. The amount of implant oversizing is crucial in this context: too much oversizing makes seating of the implant difficult, requires high forces during implantation – which might cause fractures or fissures of the bone, and results in small contact areas between implant and bone [29]. Too little oversizing might result in a condition, in which the implant can move with respect to the bone, especially during situations with low compressive forces in the joint. Such a situation on the cup side is, for example, the reduction of the joint after implantation of the implant components, when the head is moved over the rim of the cup.

2.4
Tissue Damage and Joint Tensioning

The amount of soft tissue damage during surgery and the tension in the remaining soft tissue after THA implantation are also important factors for the stability of the joint. It has been shown that the surgical approach influences the dislocation rate [6, 19]. Other studies, however, investigated the function of the hip joint after different approaches and did not show a functional difference [26]. The controversy might be due to the complex situation: The surgical approach by itself determines which muscle groups have to be cut or detached or split by the surgeon. The surgeon, however, determines the extent of the involvement of the soft tissue structures. Furthermore, the positioning of the implant components, in order to reproduce the anatomical joint centre and offset, heavily influences the biomechanical situation in the joint during loading. This might explain why two different surgeons can achieve different results with respect to dislocation rate with the same surgical approach.

Hard tissue damage can also occur during implantation. This can either result in direct complete fractures or fissures of the bone, or in micro-fractures of trabecular bone, which can develop to a complete fracture later on (Fig. 2.5).

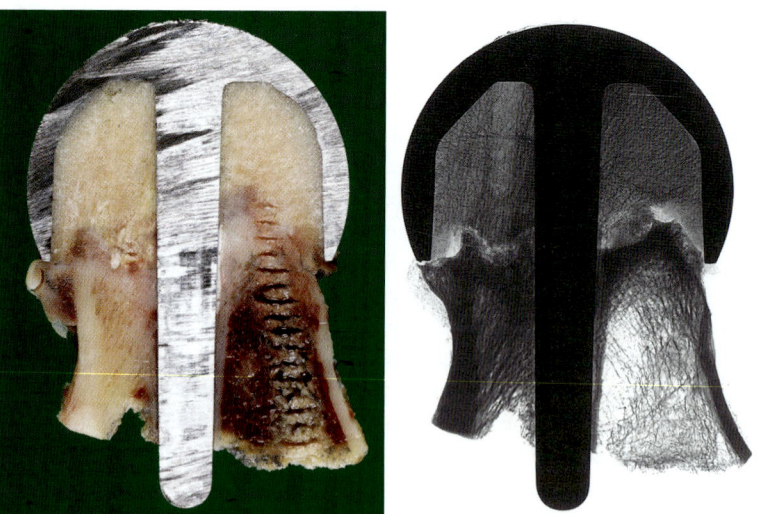

Fig. 2.5 Histological section and contact X-ray of a revised resurfacing prosthesis (1,374 days after implantation). The fracture gap across the femur-head junction demonstrated pseudarthrotic tissue. The fracture might have been initiated by high impaction forces during surgery, which were required to seat the implant due to the massive amount of cement used (preparation and section courtesy of Drs. Joseph Zustin and Michael Hahn, University Hospital Hamburg)

2.5
Component Orientation

Component orientation and position is probably the most important biomechanical aspect for the tribological and functional success of a THA procedure. The material and manufacturing issues, which have been the limiting factor for the success of the procedure in the past, have been successfully addressed in the last two or three decades.

On the acetabular side, poor component position directly influences friction, wear, and the risk of dislocation due to the reduced effective jumping distance [18, 27, 28]. In metal-on-metal articulations, run-away wear (Fig. 2.6a) with all the possible biological implications such as metallosis or pseudo-tumors can be the consequence. Malpositioning in large metal-on-metal bearings usually involves cup inclinations above 50° and/or anteversion above 15° (Table 2.2). In Ceramic-on-ceramic bearings, rim loading can cause stripe wear,

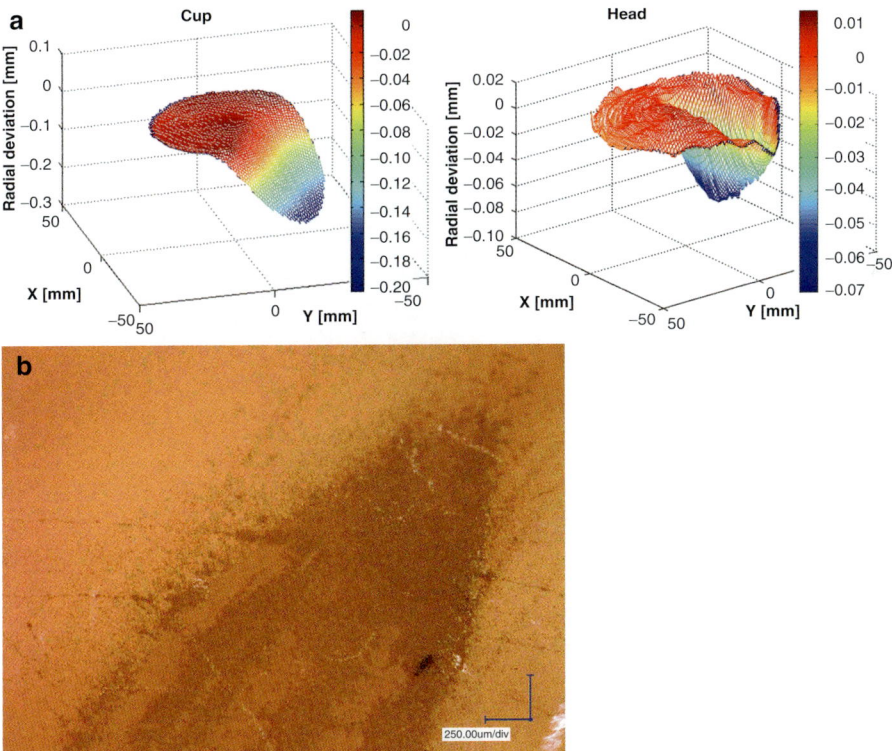

Fig. 2.6 (**a**) Example for the deviation from spherical for a rim-loaded metal-on-metal implant couple. The cup shows a distinct rim-loading pattern with wear dominantly at the edge of the cup. The head shows a deep scar at the location, where it was articulating with the rim (1,344 days in situ; Cup inclination: 57°, cup anteversion: 27°; total wear: head 34.4 mm³, cup: 68.9 mm³; revision reason: metallosis). (**b**) SEM picture of the stripe wear on a retrieved Al_2O_3 ceramic ball head (*dark area*)

Table 2.2 Time in situ and wear rate for revised hip resurfacing prostheses and XL-metal heads and cups. Cup inclination and anteversion were determined from ap-X rays using IMATRI public domain software (www.imatri.net). Classification into the rim loaded group was based on the wear pattern. Parameters are reported for implants exhibiting rim load pattern and implants not exhibiting this pattern (*Oneway Analysis of Variance*). For couples (*head and cup*), the wear rate of the head is reported

Rim loading	Time in situ (days)			Cup inclination (°)			Cup anteversion (°)			Wear rate (mm³/year)		
	n	Mean	SD	n	Mean	SD	n	Mean	SD	n	Mean	SD
No	145	243	327	114	48.1	8.2	106	10.2	8.0	138	0.40	0.80
Yes	28	690	406	23	54.8	12.4	23	21.5	12.2	27	7.42	5.17
	$p<0.001$			$p=0.002$			$p<0.001$			$p<0.001$		

resulting in increased friction (Fig. 2.6b). Stripe wear is a local damage of the ceramic surface due to break out of ceramic grains and consequent roughening. Increased friction can result in the excitation of vibrations of the prosthesis components, which can lead to audible noise phenomena, if the friction is large enough [30]. This phenomenon is mostly observed for hard-on-hard bearings, since hard-on-soft bearings cannot reach the required high friction coefficient. Increased friction also causes high moments at the cup-bone and head-prosthesis-bone interface, which can lead to problems with cup and stem fixation or also to problems in the fixation of the head on the stem [11, 17]. Hard-on-hard bearings are more sensitive to positioning than hard-on-soft bearings since their superior tribological characteristics rely on fluid film lubrication. If the fluid film breaks down, the tribological characteristics deteriorate rapidly.

On the femoral side, component position is becoming a more and more important issue since shorter prostheses are becoming more popular due to their bone conserving philosophy. It has to be realized that shorter prostheses also have a shorter lever arm to resist the moments introduced by the joint force: the shorter the stem, the higher the loading of the interface between the stem and the bone [7]. Since shorter stems typically also have less surface interface with the bone, the local bone stresses are even higher. This stress rising consequence of short stem designs mainly plays a role in uncemented stems during the ingrowth phase. If the lever arm of the joint force exceeds the load capacity of the anchoring bone, the prosthesis can rotate into varus (Fig. 2.7). In the

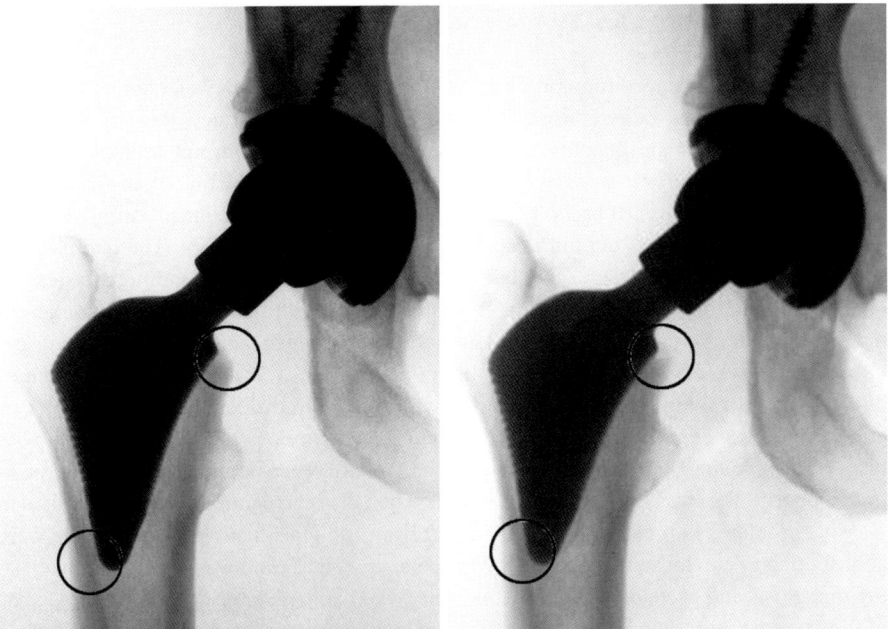

Fig. 2.7 A short hip endoprosthesis direct post operatively (*left*) and 6 month post op (*right*). Due to the very large head length and the small prosthesis size the implant rotated slightly (3°) into varus (*circles* indicate areas, where the migration can be seen)

2

worst case, fracture of the calcar can be the result. Once the bone has ingrown, this problem is greatly reduced since the ingrown bone-implant interface can also transfer tension.

2.6
Final Remarks

Some surgeons call total hip arthroplasty the most successful surgery in the history of orthopedics. This certainly seems justified considering the growing number of surgeries performed every year and the success rates in the registries. From a biomechanical point of view, the problem of THA is under control, as long as the patient and surgeon act carefully and responsibly. Established implants and bearing materials have clinically been shown to be successful in a vast majority of patients over periods in excess of 15 years. However, certain developments have to be watched carefully in order to maintain this success story:

- The continuous development of shorter prostheses (increasing the risk of bone failure)
- Lower wearing bearing materials (developed in the simulator under optimal conditions)
- New smaller surgical approaches (making the positioning of the implants more difficult)
- Surgeons doing only a few THA cases per year

There is a continuing need to improve implants and utilize newly available materials, but this process has to carefully consider the risks and side effects of new developments and not purely focus on the benefits. Continuous surgeon education and training for new implants and procedures is an essential requirement for the introduction of any new developments into clinical use. It has to be realized that successful pre-clinical testing does not guarantee clinical success but rather comprises a minimal requirement. The international standards should be extended to include testing of adverse implant conditions instead of only considering an optimal situation. The story of the extra-large heads for metal-on-metal articulations are a good example: Larger heads do reduce the risk for dislocation and the amount of wear debris generation (in the simulator). This was the reason for their introduction. The advantage of extra-large heads with respect to dislocation remains, to a certain extent, even if the implants are positioned poorly. In contrast, the advantages with respect to wear and friction are not just reduced, but are reversed if the implants are positioned poorly. This was not realized at the time when they were introduced, and resulted in massive problems in their clinical application. High friction situations in the patient were contributing to cup loosening and taper problems, as well as run-away bearing wear, causing metallosis and pseudo-tumors. The patients, who had high expectations for these state of the art prostheses, nourished by industry and surgeons, were greatly disappointed when they had to realize the unexpected problems. A THA system, which has been optimized from a biomechanical and tribological point of view, but requires an accuracy of positioning and assembly, which cannot be routinely achieved by an experienced user, is probably not the best option for the patient.

References

1. Amlie, E., Hovik, O., Reikeras, O.: Dislocation after total hip arthroplasty with 28 and 32-mm femoral head. J. Orthop. Traumatol. **11**, 111–115 (2010)
2. Archbold, H.A., Slomczykowski, M., Crone, M., Eckman, K., Jaramaz, B., Beverland, D.E.: The relationship of the orientation of the transverse acetabular ligament and acetabular labrum to the suggested safe zones of cup positioning in total hip arthroplasty. Hip Int. **18**, 1–6 (2008)
3. Australian Orthopaedic Association: National Joint Replacement Registry Annual Report (2010)
4. Bergmann, G., Deuretzbacher, G., Heller, M., Graichen, F., Rohlmann, A., Strauss, J., Duda, G.N.: Hip contact forces and gait patterns from routine activities. J. Biomech. **34**, 859–871 (2001)
5. Bergmann, G., Graichen, F., Rohlmann, A., Bender, A., Heinlein, B., Duda, G.N., Heller, M.O., Morlock, M.M.: Realistic loads for testing hip implants. Biomed. Mater. Eng. **20**, 65–75 (2010)
6. Berry, D.J., von Knoch, M., Schleck, C.D., Harmsen, W.S.: Effect of femoral head diameter and operative approach on risk of dislocation after primary total hip arthroplasty. J. Bone Joint Surg. Am. **87**, 2456–2463 (2005)
7. Bishop, N.E., Burton, A., Maheson, M., Morlock, M.M.: Biomechanics of short hip endoprostheses – the risk of bone failure increases with decreasing implant size. Clin. Biomech. (Bristol. Avon.) **25**, 666–674 (2010)
8. Bishop, N.E., Waldow, F., Morlock, M.M.: Friction moments of large metal-on-metal hip joint bearings and other modern designs. Med. Eng. Phys. **30**, 1057–1064 (2008)
9. De, H.R., Pattyn, C., Gill, H.S., Murray, D.W., Campbell, P.A., De, S.K.: Correlation between inclination of the acetabular component and metal ion levels in metal-on-metal hip resurfacing replacement. J. Bone Joint Surg. Br. **90**, 1291–1297 (2008)
10. Fottner, A., Utzschneider, S., Mazoochian, F., von Schulze, P.C., Jansson, V.: Cementing techniques in hip arthroplasty: an overview. Z. Orthop. Unfall. **148**, 168–173 (2010)
11. Garbuz, D.S., Tanzer, M., Greidanus, N.V., Masri, B.A., Duncan, C.P.: The John Charnley Award: metal-on-metal hip resurfacing versus large-diameter head metal-on-metal total hip arthroplasty: a randomized clinical trial. Clin. Orthop. Relat. Res. **468**, 318–325 (2010)
12. Griffin, W.L., Nanson, C.J., Springer, B.D., Davies, M.A., Fehring, T.K.: Reduced articular surface of one-piece cups: a cause of runaway wear and early failure. Clin. Orthop. Relat. Res. **468**, 2328–2332 (2010)
13. Hay, J.: The Biomechanics of Sports Technique. Prentice-Hall, Englewood Cliffs (1978)
14. Johnston, R., Smidt, G.: Hip motion measurements for selected activities of daily living. Clin. Orthop. Relat. Res. **72**, 205–215 (1970)
15. Kärrholm, J., Garellick, G., Herberts, P., Rogmarck, C.: Swedish Hip Arthroplasty Register 2008. Department of Orthopaedics. Sahlgrenska University Hospital, Göteborg (2009)
16. Kluess, D., Zietz, C., Lindner, T., Mittelmeier, W., Schmitz, K.P., Bader, R.: Limited range of motion of hip resurfacing arthroplasty due to unfavorable ratio of prosthetic head size and femoral neck diameter. Acta Orthop. **79**, 748–754 (2008)
17. Long, W.T., Dastane, M., Harris, M.J., Wan, Z., Dorr, L.D.: Failure of the Durom Metasul acetabular component. Clin. Orthop. Relat. Res. **468**, 400–405 (2010)
18. Morlock, M.M., Bishop, N., Zustin, J., Hahn, M., Ruther, W., Amling, M.: Modes of implant failure after hip resurfacing: morphological and wear analysis of 267 retrieval specimens. J. Bone Joint Surg. Am. **90**(Suppl 3), 89–95 (2008)
19. Muller, M., Tohtz, S., Springer, I., Dewey, M., Perka, C.: Randomized controlled trial of abductor muscle damage in relation to the surgical approach for primary total hip replacement: minimally invasive anterolateral versus modified direct lateral approach. Arch. Orthop. Trauma Surg. **131**(2), 179–189 (2011)
20. Nassutt, R., Wimmer, M.A., Schneider, E., Morlock, M.M.: The influence of resting periods on friction in the artificial hip. Clin. Orthop. Relat. Res. **407**, 127–138 (2003)

21. National Joint Registry for England and Wales: 7th Annual Report 2009. Hemel Hempstead, U.K. (2010)
22. Oehy, J., Bider, K.: Design parameter to improve range of motion (ROM) in total hip arthroplasty. In: Lazennec, J.Y., Dietrich, M. (eds.) Bioceramics in Joint Arthroplasty, pp. 149–156. Steinkopf, Darmstadt (2004)
23. Padgett, D.E., Lipman, J., Robie, B., Nestor, B.J.: Influence of total hip design on dislocation: a computer model and clinical analysis. Clin. Orthop. Relat. Res. **447**, 48–52 (2006)
24. Pauwels, F.: Der Schenkelhalsbruch – ein mechanisches Problem. Ferdinand Enke, Stuttgart (1935)
25. Peters, C.L., McPherson, E., Jackson, J.D., Erickson, J.A.: Reduction in early dislocation rate with large-diameter femoral heads in primary total hip arthroplasty. J. Arthroplasty **22**, 140–144 (2007)
26. Pospischill, M., Kranzl, A., Attwenger, B., Knahr, K.: Minimally invasive compared with traditional transgluteal approach for total hip arthroplasty: a comparative gait analysis. J. Bone Joint Surg. Am. **92**, 328–337 (2010)
27. Sariali, E., Lazennec, J.Y., Khiami, F., Catonne, Y.: Mathematical evaluation of jumping distance in total hip arthroplasty: influence of abduction angle, femoral head offset, and head diameter. Acta Orthop. **80**, 277–282 (2009)
28. Sariali, E., Stewart, T., Jin, Z., Fisher, J.: In vitro investigation of friction under edge-loading conditions for ceramic-on-ceramic total hip prosthesis. J. Orthop. Res. **28**, 979–985 (2010)
29. Spears, I.R., Pfleiderer, M., Schneider, E., Hille, E., Bergmann, G., Morlock, M.M.: Interfacial conditions between a press-fit acetabular cup and bone during daily activities: implications for achieving bone in-growth. J. Biomech. **33**, 1471–1477 (2000)
30. Weiss, C., Gdaniec, P., Hoffmann, N.P., Hothan, A., Huber, G., Morlock, M.M.: Squeak in hip endoprosthesis systems: an experimental study and a numerical technique to analyze design variants. Med. Eng. Phys. **32**, 604–609 (2010)
31. Wolff, J.: Über die innere Architektur der Knochen und ihre Bedeutung für die Frage von Knochenwachsthum. Virchows Arch. A Pathol. Anat. Histopathol. **50**, 389–450 (1870)

Ceramic Hip Replacements: Wear Behavior Affects the Outcome – A Tribological and Clinical Approach

3

Meinhard Kuntz, Sylvia Usbeck, Thomas Pandorf, and Ricardo Heros

3.1
Introduction

Recently, the endoprosthetic treatment of younger and more active patients has increased and got into focus of the orthopedic community. This patient group wants to live an active life, posing increasing demands on the mechanical and tribological properties of artificial hip implants. In addition, life expectancy and activity levels of older patients have been increasing [52, 81, 82, 89]. As a consequence, also the virtual picture of the old, less active patient has to be revised. In a clinical study, Wollmerstedt et al. [52, 81, 82] have measured the mean daily movement of patients (mean age 70 years) with total hip arthroplasty. They observed a mean number of two million load cycles per year which is double than those being the basis for simulator studies (one million cycles per year). On the other hand, an increased level of activity should not increase the risk of a wear induced osteolysis. The authors conclude that the bearing material plays an even more important role than has been acknowledged up to now.

The wear induced aseptic loosening is still one of the main indications for a revision surgery in THA [75, 78, 83–86]. Facing the problems being created by wear particles in metal or polyethylene bearings has to be the first aim to minimize the creation of wear debris and to avoid the resulting complications. Polyethylene particles are responsible for a large number of periprosthetic osteolyses in total joint arthroplasty [34, 83–85, 90]. Considering

M. Kuntz (✉)
Service Center Development, CeramTec GmbH, CeramTec Platz 1–9 73207,
Plochingen, Germany
e-mail: m.kuntz@ceramtec.de

S. Usbeck and T. Pandorf
Medical Products Division, CeramTec GmbH, CeramTec Platz 1–9 73207,
Plochingen, Germany
e-mail: s.usbeck@ceramtec.de; medical.products@ceramtec.de

R. Heros
Medical Products Division, CeramTec North America, 533 Princeton Cove Memphis,
TN 38117, USA
e-mail: rheros@ceramtec.com

K. Knahr (ed.), *Tribology in Total Hip Arthroplasty*,
DOI: 10.1007/978-3-642-19429-0_3, © 2011 EFORT

shells and inserts made of polyethylene, influencing factors are, among others, the size of the femoral ball head [91–93] and the activity level of the patient [15, 76]. It is common consensus that large bearing diameters due to their enhanced range of motion (ROM) are preventing subluxation, dislocation, and impingement, and are positively influencing joint stability. Postoperative dislocation rates of up to 10% [87, 88] visualize that the dislocation problem is still a major concern, and that implant stability together with a sufficient ROM is an important success factor in THA [51]. Nevertheless, a larger diameter of the bearing couple in connection with polyethylene inserts leads to an increased wear volume [91–93].

For ceramic bearing couples, the problem of wear volume is more or less solved. Clinical experience states that they show the lowest particle emission and osteolytic potential of all bearing materials in use. As a consequence, ceramic implants are well suited for risk patients, e.g., in the case of metal allergy [65]. Furthermore, ceramic bearing couples show no increase of wear volume with increased bearing diameter [94]. The above mentioned problems have lead to a more widespread use of ceramic bearing couples.

The first applications of ceramic bearing couples in orthopedics were made of pure alumina (Al_2O_3) [1, 3, 17, 32]. Since 1971, they are used in this area, and more than five million components have been implanted [16, 33, 42]. Enhancements regarding the microstructure and the reliability of ceramic materials have been reached mainly by improving the production process, leading to a significant increase in mechanical strength. Composite ceramics with even more improved mechanical properties offer new areas of application such as larger bearing couples for better range of motion and joint stability. In the following, the main characteristics as well as the mechanical behavior of a special ceramic composite material are analyzed. Its clinical potential is assessed as being very promising [19].

3.2
Ceramic Selection for Bearing Couples

Pure alumina ceramics have been the standard for more than 40 years for hip arthroplasty due to their superior wear performance and biocompatibility. While at the beginning of the use of alumina components the fracture rate was comparatively high, improvements in ceramic technology, quality management and design optimization has led to highly reliable application of alumina and very low fracture rate (today ~0.02%). Today, the material properties which are in a physically realistic expectation for alumina are almost reached.

There are mainly three aspects in the current state of the art in arthroplasty which push the need for a higher performance material than can be provided by pure alumina:

- A strong tendency to use larger wear couples for improved range of motion and comfort for the patient
- Higher reliability in the case of unforeseen severe impact on the ceramic components, including disadvantageous wear conditions
- Increasing need for the use of ceramics for other artificial joints like knee and spine, particularly of interest to consequently avoid any possible implication of metal debris

The ceramic material providing the highest strength is *zirconia* due to the unique transformation toughening effect (as explained later). High performance yttria stabilized zirconia (Y-TZP) can provide more than double the strength in comparison to pure alumina.

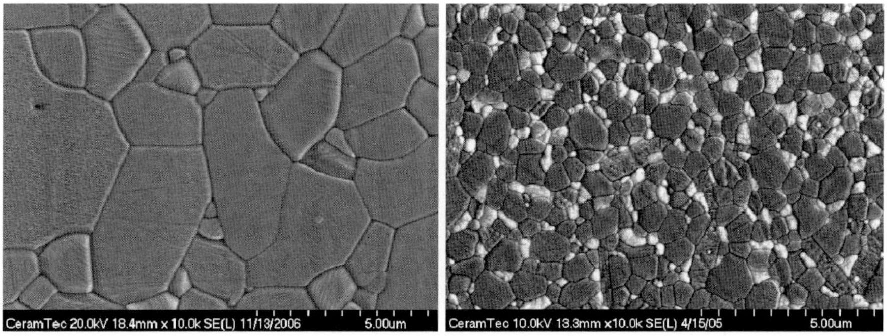

Fig. 3.1 Microstructure of pure alumina BIOLOX®*forte* and ZTA BIOLOX®*delta*. Note that the magnification in both figures is identical. Alumina appears gray, zirconia white. Zirconia content in BIOLOX®*delta* is 17vol.%

However, the use of Y-TZP for wear applications has almost disappeared in Europe and USA due to the particular problem of hydrothermal aging, i.e., undesired phase transformation in human body environment with adverse impact on the surface integrity of the component. Moreover, zirconia shows a significantly lower hardness and thermal conductivity than alumina which is also considered a disadvantage for Y-TZP.

An ideal ceramic material for arthroplasty should combine the advantages of zirconia and alumina (BIOLOX®*forte*, CeramTec GmbH, Plochingen, Germany), i.e., the high strength and toughness on the one hand, and high hardness and inertness on the other hand. This concept can be achieved using a composite of these ingredients, i.e., zirconia toughened alumina (ZTA). Such a material is available in the market since 2002, under the trade name BIOLOX®*delta* (CeramTec GmbH, Plochingen, Germany). It was just in the year 2009 that the total number of hip components produced with ZTA almost balanced the number of those with the established pure alumina. It is expected that ZTA will dominate the near future of bioceramics for arthroplasty.

Figure 3.1 shows a comparison of the microstructures of alumina and ZTA as they represent the current state of the art.

3.3
Benefit of the Phase Transformation of Zirconia

The benefit in crack resistance which is obtained from incorporating zirconia into an alumina matrix as shown in Fig. 3.2 is well known in the science of high performance ceramics.

Figure 3.2 represents a realistic part of the microstructure. The gray particles refer to the alumina matrix, yellow to tetragonal zirconia. The phase transformation of zirconia is indicated by the change to red color. In the case of severe overloading, crack initiation and crack extension will occur. High tensile stresses in the vicinity of the crack tip trigger the tetragonal to monoclinic phase transformation of the zirconia particles. The accompanied volume expansion of approx. 4% leads to the formation of compressive stresses which are efficient for blocking the crack extension [18, 21].

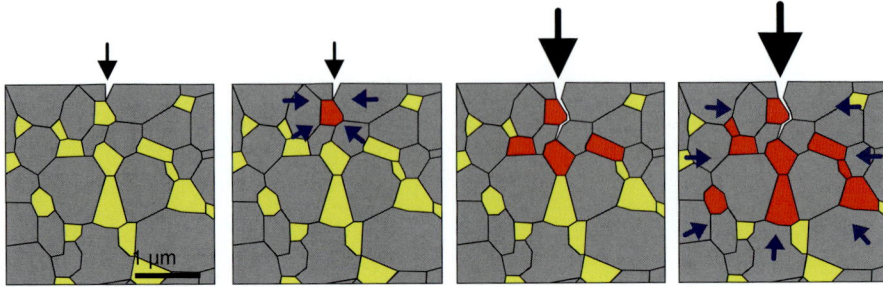

Fig. 3.2 Reinforcing mechanism in BIOLOX®delta at crack initiation and propagation. Yellow particles represent tetragonal zirconia. Color change to red indicates monoclinic phase transformation. Arrows show the region of compressive stresses due to phase transformation

As it is demonstrated in Fig. 3.2, this reinforcing mechanism is fully activated within a region of a few micrometers. For the macroscopic performance of the material it is very important that immediately at the beginning of crack initiation the reinforcing mechanisms are also activated. Regarding this mechanism, one should keep in mind that the average distance between the reinforcing zirconia particles is approx. 0.3 µm, i.e., similar to the grain size. Thus, the reinforcement is activated immediately when any microcrack is initiated. This is of particular interest for the significant advantage of ZTA (BIOLOX®*delta*) under severe wear conditions.

3.4
Simulation of Severe Wear Conditions

Under normal conditions, the wear of ceramic surfaces during the life time of a patient is almost negligible. The amount of wear debris in comparison to other materials is reduced by orders of magnitude. Moreover, zirconia and alumina are bioinert, thus no detrimental effects are expected from the debris.

However, under adverse circumstances the wear conditions can be disturbed. In particular, it is possible that the ball head leaves the ideal position inside the insert due to impingement or microseparation. In this case, the wear conditions shift from surface contact to point contact which lead to highly located compressive stresses. Such a situation is the main reason for stripe wear, i.e., a clearly distinguished area with significantly higher surface erosion.

In order to show the performance of ZTA vs. pure alumina at high impact wear, several experiments have been performed [79, 80]. In Fig. 3.3 (*left*), the configuration of a microseparation test as performed by Stewart et al. [80] is shown.

It is well known, that in the vicinity of highly located contact the surface can be damaged due to microcracking, as schematically depicted in Fig. 3.3 (*right*). This is most likely the reason for increased surface erosion under stripe wear conditions. On the other hand, as it was discussed under Fig. 3.2, it should be expected that a material with an efficient toughening mechanism in the micron-scale shows an improved performance under stripe wear conditions. This expectation is convincingly shown from the experiments performed by Stewart et al. [80], see Fig. 3.4.

Fig. 3.3 *Left*: Simulation of stripe wear conditions, ball head leaves center position *Right*: Schematic description of surface microcracking in vicinity of high contact load

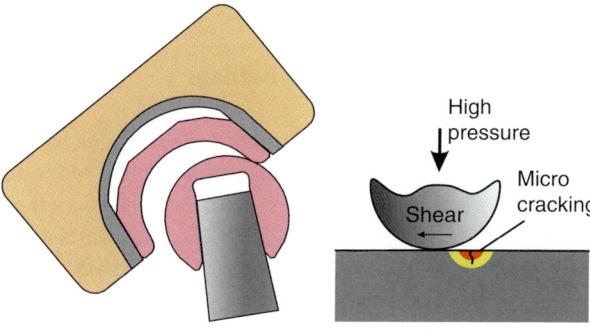

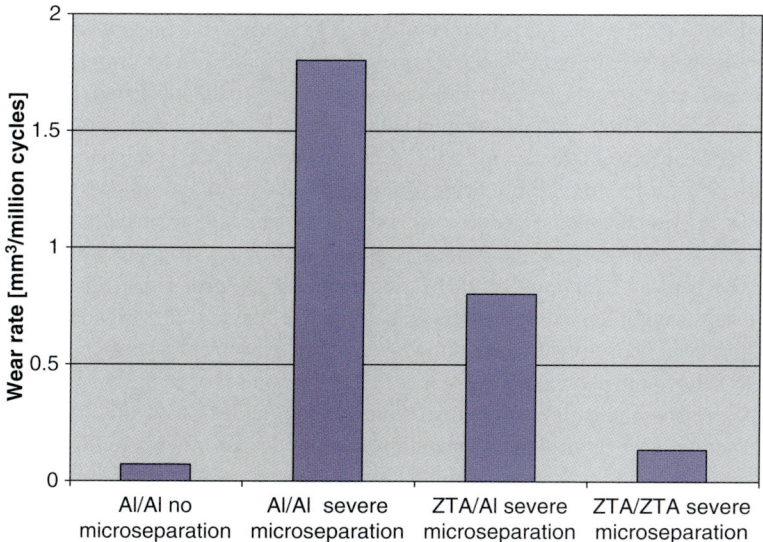

Fig. 3.4 Results of microseparation tests after [*Stewart*], average of five million cycles

In this experiment, three different couples of ceramics are used: (1 and 2) pure alumina vs. alumina, (3) ZTA vs. alumina and (4) ZTA vs. ZTA. As a reference, the pure alumina couple was also tested without microseparation. As expected, the standard wear test without microseparation revealed extremely low wear rate of < 0.1 mm³/million cycles. The same couple, tested under extreme conditions of five million cycles microseparation, shows more than one order of magnitude higher wear rate due to contact load damage. It should be considered that these testing conditions are unrealistically severe. As the most important result of this experiment, the ZTA couple (4) performs excellent even after five million cycles of microseparation. The total wear is only marginally higher than the wear of the pure alumina under optimal conditions. This result supports the hypothesis that ZTA materials provide higher reliability under disadvantageous wear conditions. The performance is clearly correlated to the intrinsic reinforcement of the zirconia which efficiently prevents microcracking.

3.5
Tribological Aspects of Ceramic Bearings

Ceramic-on-ceramic bearings made of pure alumina have a long clinical history [42, 43]. The performance in the early years was sometimes not satisfying which was due to material, design, and surgical related factors. Over the years, the improvements in materials and manufacturing as well as the adaptation of design and surgical technique have led to a variety of safe and successful designs. Nevertheless, the potential of ceramic-on-ceramic bearing couples to reduce the risk of osteolysis is based on their excellent wear behavior [14, 16] as well as their high degree of biocompatibility which is caused by the minimal risk of ionization of ceramic particles [6]. The long-term in-vitro volumetric wear rate for well-functioning ceramic-on-ceramic bearings made of pure alumina has been measured as varying from 0.02 mm^3/mc [35] to 0.04 mm^3/mc [13], whereas the linear wear rate for well-functioning in-vivo pairings has been reported between 0.025 μm to 5 μm [28, 38, 45].

The request for larger bearings arising from the increased patient demands has led to bearing diameters that require special attention with respect to frictional moments and wear [10]. In [36], the wear behavior of ceramic bearing couples up to 44 mm is compared to that of the clinically well-functioning 28 mm made of pure alumina. It could be shown that even large deviations in clearance and roundness do not lead to excessive wear but to a steady-state wear rate in the same range as the 28 mm pure alumina bearing. In [2], the friction moments for different hard-on-hard bearings and bearing diameters have been investigated. It could be shown that assuming same diameters by linear scaling of the resulting friction moments, the ceramic-on-ceramic bearing exhibited the lowest friction moment, therefore, promising the best wear behavior in clinical use compared to metal-on-metal or ceramic-on-metal.

Stripe wear is a phenomenon known from hard-on-hard bearings [8, 49]. The occurrence of stripe wear in vivo may depend on implant design, orientation in the hip joint, patient activities and implant duration [31, 49]. Although only anecdotal reports exist for a correlation between ceramic wear debris and osteolysis [48], the phenomenon and its reproduction in-vitro has been studied. In [9], the shape and surface roughness of in-vitro created stripe wear in the ZTA ceramic BIOLOX®*delta* has been compared with the stripe wear of retrieved parts of the same material. The reasons for revision were non wear-related. Results show that the artificially reproduced stripe wear shows good correlation with that from the explanted parts.

3.6
Clinical Performance and Their Relation to the Material of the Bearing Couple

Wear and its consequences are limiting factors of total hip arthroplasty in young and active populations. Aseptic loosening due to polyethylene wear remains the most common case of implant failure [83–86, 90]. In the Swedish National Hip Register, aseptic loosening accounts for 75% of all hip revisions [78]. The Finnish Arthroplasty Register suggested that the limiting factors for survival of total hip arthroplasty for patients younger than 55 years old was polyethylene wear [75]. Schmalzried et al. [76] demonstrated that polyethylene wear was related to activity and activity was related to younger patients' age. The life expectancy and activity demands are increased. This results in up to a ten-fold increase in the tribological demands of hip implants [77].

It is clear from the literature that current orthopedic surgical practice is in need of further alternatives in order to complement the metal against ultra high molecular weight or highly crossed linked polyethylene wear couple for a category of patients defined as younger and more active.

Several reports demonstrated that the use of ceramic implants may contribute to a large extent to diminishing wear debris and, hence, to a significant reduction of the revision rate in THA. The use of ceramic-on-polyethylene bearing couples has enabled a two to fivefold reduction in the revision rate compared to metal on polyethylene bearing couples [7, 53, 59, 69, 70].

Alumina ceramic on ceramic bearings have shown a 100-fold decrease in wear compared to highly cross-linked polyethylene materials [54].

Cell culture studies demonstrated that ceramic wear particles are more biocompatible and have the lowest functional biological activity and osteolytic potential compared to other bearing materials [54]. Baldini et al. [55] investigated alumina ceramic-on-ceramic hip replacements (BIOLOX® 32 mm, BIOLOX®*forte* 28 mm). The average service life of the explants had been 8 (1–17) years. The investigation showed extraordinarily minimal wear rates and no negative biological reactions to the released ceramic particles. Wear debris particles were largely absent. In the few cases particles were observed, however their volume was minimal. No evidence of foreign body reaction or inflammation was observed. No cases of extensive osteolysis or cytotoxic effects were found. These results offer additional confirmation of results obtained from earlier investigations conducted by various work groups.

Extremely low wear has also been recorded in hip simulator studies and in vivo for alumina ceramic on ceramic bearings. These results are reflected in clinical results where there has been a very low incidence of osteolysis.

3.6.1
North American Experience

D'Antonio and Capello [56] presented a review of the 52,000 ceramic-on-ceramic hips implanted since the 2003 FDA approval of Stryker ceramic hip arthroplasty products. They reported that a total of four insert fractures (0.008%) and nine femoral head fractures (0.017%) were seen in the group. Further data comparing metal-on-polyethylene and ceramic-on-ceramic couples at greater than 7 years follow-up found a 7.5% revision rate in the control group (metal-on-polyethylene) versus a 2.7% revision rate for ceramic-on-ceramic. There were no fractures in the ceramic cohort, while the control group experienced osteolysis and revisions attributable to wear [99]. Capello et al. [56] reported on 475 hips at an average 8 years of follow-up. Cortical erosions were seen less in the ceramic-on-ceramic hips (1.4%) than in the metal-on-polyethylene group (30.5%), and there was no aseptic loosening in the ceramic group. Overall, there was less than 1% requiring revision in the ceramic-on-ceramic hips.

Murphy et al. [57], in a report on the Wright MT IDE trials, included 1,709 hips in 1,484 patients across 12 centers. At an average 8-year follow up, there were four ceramic liner fractures including intraoperative chips (0.27%) and two re-operations for instability. While regulatory approvals have historically forced ceramics to introduce new sizes at a slower pace than metal components with respect to head size, Murphy concluded that the relative lack of neck and liner options did not impede safe surgery with these components.

In another study, Murphy et al. [30] treated 360 patients (mean age of 51.7 ± 12.3 years) with 418 ceramic-on-ceramic bearing couples. 41 of these cases (11%) were revisions. No case of osteolysis could be observed. The authors concluded that the results of these studies are very promising due to the young age of the patients and the high incidence of revision cases.

Mesko et al. [58] reported on a comparison of alumina ceramic-on-ceramic (BIOLOX®*forte*) and metal-on-polyethylene bearing couples in patients whose activity levels are well above average. The 10-year survival rate was 96.8% for the ceramic-on-ceramic group and only 92% for the metal-on-polyethylene group. Mesko concluded that the long-term safety of ceramic-on-ceramic bearings is demonstrated by the low incidence of revision in comparison to metal-on-polyethylene.

Steppacher et al. [46] reported on first results of a prospective study in which 123 dysplastic hips (Crowe type I and II) in 108 patients (mean age of 47.6 ± 12.7 years) have been treated with a ceramic-on-ceramic bearing couple (alumina, BIOLOX®*forte*, 28 mm, 32 mm). No cases of osteolysis or dislocation have been observed. The authors conclude that these results are very promising in young patients with dysplasia after 2–10 years follow-up.

Lewis et al. [25] reported on first results of a prospective, randomized medium- to long-term study in which ceramic-on-ceramic and ceramic-on-polyethylene bearing couples were compared in a follow-up time of up to 10 years. 55 active patients with a mean age of 42.2 years received 56 cementless components. 30 implantations were made with the hard-on-hard bearing couple, 26 with the hard-on-soft bearing couple. In all hips, 28 mm femoral ball heads were used. Signs of wear debris were identified in 25 of the 26 ceramic-on-polyethylene hips, but only in 12 of the 23 ceramic-on-ceramic hips. The linear wear rate per year was significantly lower for the ceramic-on-ceramic group (0.02 mm) than for the ceramic-on-polyethylene group (0.11 mm). The authors concluded that the ceramic bearing couple is a safe long-term option avoiding possible risks that might arise through the use of polyethylene or metal components.

The review of the experience in the American environment clearly shows that there is a significant reduction in wear achieved when the ceramic on ceramic articulation is used in a challenging patient group, the younger and more active patient. The use of this articulation in large numbers in the USA is now approaching the 10 year level with very significant reductions in osteolysis.

3.6.2
European Experience

The reported long-term clinical and radiographic results with alumina ceramic-on-polyethylene and ceramic-on-ceramic demonstrated the value of ceramics as a femoral bearing surface in THA.

Ihle et al. [59] reported on significantly lower wear rate and debris in ceramic-on-polyethylene bearing couple, less osteolysis and revisions when compared to metal-on-polyethylene bearing couple after 20 years in vivo. The mean age of the patients was 52 years. They observed no ceramic fractures. The average annual wear rate was 0.107 mm for the ceramic femoral ball head group and 0.190 mm for the metal femoral ball head group. At 13.8%, the revision rate for the ceramic femoral ball head group was significantly lower than the 46.2% revision rate for the metal femoral ball head group.

Zichner et al. [53] reported that the revision rate in hard-on-soft bearing couples after 10 years was five times higher with metal femoral ball heads than with ceramic femoral ball heads. In this study, the same THA systems using the same surgical technique were implanted. This result was supported in newer investigations by Kusaba et al. [22] and Dahl et al. [5].

In a prospective study, Dahl et al. [5] reported a significantly increased wear with metal femoral ball heads (CoCr, 28 mm) compared to ceramic femoral ball heads (alumina, BIOLOX®*forte*, 28 mm). The wear rates of the control group correlated with values from the literature. The linear wear rate of cemented polyethylene inserts which have been implanted together with ceramic femoral ball heads in 47 cases and with metal femoral ball heads in 40 cases has been investigated. All patients have been operated by one surgeon using the same surgery technique. Using radiostereometric analysis (RSA), it was observed that the linear wear in the group of the metal femoral ball heads was more than double compared to the linear wear in the group of the ceramic femoral ball heads (0.93 mm compared to 0.43 mm, $p = 0.001$).

Descamps et al. [60] presented the 15-year results of a prospective, randomized study in which the wear rates of 37 alumina ceramic-on-polyethylene THA and 37 metal-on-polyethylene THA were compared. A 28 mm femoral ball head (BIOLOX® *forte*) was used. The wear rate for ceramic-on-polyethylene (0.058 mm/year linear, 35.7 mm^3/year volumetric) was significantly lower than those for metal-on-polyethylene (0.102 mm/year linear, 62.8 mm^3/year volumetric). This corresponds to a reduction in head penetration of 44%. Descamps concluded that these results are comparable to results obtained in earlier studies of ceramic-on-polyethylene and metal-on-polyethylene at a follow-up of more than 10 years.

The reported European experience with an alumina-on-alumina combination showed a mean wear rate of 0.025 µm/year and limited osteolysis up to 10 years after arthroplasty [61–63]. When comparing 28 bilateral arthroplasties (one alumina ceramic-on-ceramic and the contralateral alumina ceramic-on-polyethylene) at 20-year revision-free follow-up, Hernigou et al. [64] saw more wear and osteolysis in the ceramic-on-polyethylene hip than the ceramic-on-ceramic hip.

Toni et al. [48] has compiled data on 7005 CoCr bearing couples (BIOLOX®, BIOLOX®*forte*, BIOLOX®*delta*). During the period from 2006 to 2008, 686 ceramic-on-ceramic bearing couples (BIOLOX®*delta*) were implanted. The 17-year follow-up investigation of 147 patients treated consecutively with a 32 mm alumina ceramic-on-ceramic bearing couple between 1990 and 1991 revealed no cases of osteolysis, not even among those patients whose replacements showed increased wear as a result of suboptimal positioning. Toni suggested that these results support the claim that the use of ceramic-on-ceramic couples will help to minimize the risk of osteolysis. There were no cases of noise development or fracture. He called attention to the problem of false osteolysis positives, pointing out that it is necessary to first examine the preoperative X-rays to avoid mistaking older bone defects for cases of postoperative osteolysis.

Raman et al. [66] showed that the use of large femoral ball head diameters (36 mm, 40 mm) in ceramic-on-ceramic THA (BIOLOX®*delta*) can help to lower the risk for dislocation and secure a large range of motion. He also applies it to patients older than 60 with insufficient muscular stability. A total of 319 consecutive primary THAs in 302 patients with an average age of 65 (11–82) years were clinically and radiologically examined at a follow-up of 12 months. No dislocations and no cases of aseptic cup loosening were observed.

3.6.3
Asian Experience

Recently published clinical results from Asia show that modern ceramic-on-ceramic bearing couples have low risk of osteolysis and revision rates. This is especially valid for younger and more active patients, patients with a high incidence of previous hip surgeries and patients with dysplasia. Those patient groups are in general associated with an increased risk of complication and revision due to wear induced osteolysis and instability.

Kusaba et al. [22] described hip dysplasia as one of the most common indications for hip arthroplasty in Japan. Between July 1998 and October 2008, 1078 dysplastic hips were treated with a cementless alumina ceramic-on-ceramic bearing couple (BIOLOX®*forte*). 86 hips in 79 patients (mean age of 53 years) have been followed (completed follow-up time 5 years). No dislocations or revisions were observed.

In another study Kusaba et al. [23] analyzed the surface structure of 36 explanted metal femoral ball heads (CoCr, 32 mm, average 9 years in vivo) and 56 ceramic femoral ball heads (alumina, BIOLOX®*forte*, 32 mm, average 8.5 years in vivo). The linear wear rate per year of the polyethylene inserts were lower with the ceramic femoral ball heads (0.13 mm± 0.05 mm) than with the metal ones (0.21 mm± 0.09 mm). The polyethylene wear was directly related to the roughness (Ra) of the femoral ball heads ($p < 0.05$). The average Ra value of the ceramic femoral ball heads (0.011 μm± 0.003) was lower than that of the metal femoral ball heads (0.032 μm± 0.015). Here, the roughness of the explanted ceramic femoral ball heads was comparable to that of new metal femoral ball heads. The revision rates were significantly lower with the ceramic-on-polyethylene bearing couple than with the metal-on-polyethylene bearing couple.

Kim et al. [67] evaluated alumina ceramic-on-ceramic (BIOLOX®*forte*) in 93 consecutive cementless THA. The average age at the time of surgery was 38.2 (24–45) years. No ceramic fractures were seen. Radiographs and computerized tomographic scans demonstrated no acetabular or femoral osteolysis. The survival rate with aseptic loosening as the endpoint was 100% at 11.1 years.

In another study, Kim [68] reviewed clinical and radiological results of 601 hips with alumina ceramic-on-ceramic cementless THA in 471 patients with an average age of 52.7 years at the time of surgery. The mean follow-up was 8.8 (5–12) years. No THA required revision of any component for aseptic loosening. No ceramic fractures, acetabular or femoral osteolysis were observed. Kaplan-Meier survival analysis, with aseptic loosening as the endpoint for failure, revealed a 10-year rate of survival of 100% for the acetabular and femoral component.

3.6.4
Australian Experience

Walter [98] reviewed a large series of 2503 alumina ceramic-on-ceramic THA which was performed from July 1997 to September 2004. None of these hips has required revision for osteolysis. He reported that ceramic fracture has not been a major problem with one patient requiring revision due to fracture of a ceramic insert.

Lusty et al. [26] proved in a prospective study a survival rate of 99% after 7 years. In 283 patients (mean age of 58 years), 301 hips have been treated with alumina ceramic hard-on-hard bearings (BIOLOX®*forte*). All implantations were performed in one center with identical surgical methods and identical implants. 251 patients were followed-up clinically and radiologically. Nine revisions have been performed due to periprosthetic fractures, psoas tendinitis, and femoral shortening osteotomy. Using aseptic loosening and osteolysis as the endpoint, the survival rate was 99%. The explanted inserts exhibited a low volumetric wear of 0.2 mm^3.

Lusty concluded that these results are consistent with those in other studies of alumina ceramic-on-ceramic bearings.

3.7
Current Situation on Articulation Noise

The issue of noise generation by the components implanted in total hip replacements has recently received a great deal of attention.

There are several types of noises or sounds that can be emanated from the operated hip area. These noises can be transient particularly in the early rehabilitation period as the muscles and tissues around the implants heal. In addition, there are reports of longer lasting component generated noises such as squeaking. Noises are reported from all kinds of bearings [4, 19, 37, 44] even though they are more common in hard-on-hard bearings. The percentage of noises varies greatly, from 0.6% reported by Walter [50] up to 21% published by Keurentjes [20]. This is mainly due to the different definitions of the squeaking noise given by the authors. Restrepo [39] describes it as a "high-pitched noise" whereas in [20] different types of noises are summarized, defined as a reproducible sound of squeaking, clicking, or grating, as squeaking. Toni [47] described noises as being grinding sounds likely due to third particle wear. Therefore, it is important to differentiate between the types of noises that emanate from a hip in order to examine the responsible mechanisms. Additionally, to assess the main influencing factors of the different noise types may facilitate prevention and/or treatment.

A thorough review of the case studies presented in the literature reveal a variety of complications that could cause noise generation in the so-called squeaking hips. According to [50], the occurrence of squeaking was related to a very thick anterior capsule that folded into the joint gap causing slight subluxation of the bearing. Rosneck [41] reported the possibility of femoral head edge loading on the acetabular ceramic due to high cup inclination which is also the main reason given by Lusty [27]. Although this correlation is questioned in [39], careful investigation of the herewith published data reveals that almost all hips are placed far outside the so-called Lewinnek zone [24]. Back [1] suggested a disruption of the fluid film as being the reason for squeaking in metal-on-metal resurfacing. Morlock [29] reported on a mismatch of the bearing diameter as being the cause for squeaking, whereas Eickmann [12] found metal debris in a squeaking ceramic-on-ceramic bearing couple. As can be stated from the above, all the above mentioned factors imply a frictional cause for the squeaking. Ecker [11] states that in their study all squeaking cases of ceramic-on-ceramic bearings occurred with liners with an elevated titanium alloy rim and metal contamination of the bearing. According to the findings described there, squeaking is mainly due to the materials and designs of the hip implant system including stem and shell [40] as well as component placement.

3.8
Conclusion

Medium- and long-term clinical experience with ceramic hip implants clearly states a significant reduction of revisions due to wear-related osteolysis and its complications. The use of a ceramic-on-ceramic bearing is a safe and durable option in the young and active patient avoiding the concerns of active metal ions and osteolytic polyethylene debris. In hip endoprosthetics, the new ZTA ceramic (BIOLOX®*delta*) offers the possibility of using larger femoral ball heads (larger than 36 mm) and inserts with a thinner wall thickness.

Dislocation is the second-most reason for revision in hip arthroplasty [78]. Several authors report that the risk of impingement and dislocation has been strongly reduced due to the use of 32 mm and 36 mm articulation diameters in ceramic bearing couples [71–74]. No additional wear is produced compared to a 28 mm ceramic bearing couple. In revision endoprosthetics, a special ceramic femoral ball head system (BIOLOX®OPTION) made of the mentioned ZTA-ceramic (BIOLOX®*delta*) has been developed which consists of a thin-walled ceramic femoral ball head and a metal sleeve. This system provides a secure possibility of replacing a ceramic femoral ball head in a revision surgery without the need of replacing the stem. The ball head system offers a safe solution for the rare case of ceramic component fracture and expanded applications for primary surgery [95–97].

Ongoing studies will provide further evidence of long-term outcomes and patient activity levels for ZTA-ceramic (BIOLOX®*delta*) implants.

References

1. Back, et al.: Early results of primary Birmingham hip resurfacings. JBJS (Br) **87-B**, 324–329 (2005)
2. Bishop, et al.: Friction moments of large metal-on-metal hip joint bearings and other modern designs. Med. Eng. Phys. **30**, 1057–1064 (2008)
3. Bizot, et al.: Press-fit metal-backed alumina sockets: a minimum five-year follow-up study. Clin. Orthop. Relat. Res. **379**, 134–142 (2000)
4. Bono, et al.: Severe polyethylene wear in total hip arthroplasty. Observations from retrieved AML PLUS hip implants with an ACS polyethylene liner. J. Arthroplasty **9**(2), 19–25 (1994)

5. Dahl et al.: Less wear with 28 mm aluminum oxide heads against conventional PE -A 10 year RSA study, 60th annual meeting of the Norwegian Orthopaedic Society, Oslo. 24–26 (2007)
6. Campbell, et al.: Biologic and tribologic considerations of alternative bearing surfaces. Clin. Orthop. Relat. Res. **418**(1), 98–111 (2004)
7. Capello, et al.: Ceramic-on-ceramic total hip arthroplasty: update. J. Arthroplasty **23**(7), 39–43 (2008)
8. Clarke, et al.: How do alternative bearing surfaces influence wear behavior? J. Am. Acad. Orthop. Surg. **16**, 86–93 (2008)
9. Clarke, et al.: Hip-simulator wear studies of an alumina-matrix composite (AMC) ceramic compared to retrieval studies of AMC balls with 1–7 years follow-up. Wear **267**, 702–709 (2009)
10. Dowson, et al.: A hip joint simulator study of the performance of metal-on-metal joints, simulator study of MOM hips. Part II:Design. J Arthroplasty **19**(8), Suppl.3, 124–130 (2004)
11. Ecker, T., et al.: Squeaking in total hip replacement: no cause for concern. Orthopedics **31**(9), 1–3 (2008)
12. Eickmann, et al.: Squeaking and neck-socket impingement in a ceramic total hip arthroplasty. Bioceramics **15**, 849–852 (2003)
13. Essner, et al.: Hip simulator wear comparison of metal-on-metal, ceramic-on-ceramic and crosslinked UHMWPE bearings. Wear **259**, 992–995 (2005)
14. Fisher, et al.: Long term wear of ceramic on ceramic hips. In: Lazennec, J.Y., Dietrich, M. (eds.) Bioceramics and Alternative Bearings in Joint Arthroplasty, Steinkopff-Verlag Darmstadt, 45–46 (2004)
15. Greenwald, et al.: Alternative bearing surfaces: the good, the bad, and the ugly. J. Bone Joint Surg. Am. **83**, 68–72 (2001)
16. Hamadouche, et al.: Cementless bulk alumina socket: preliminary results at six years. J. Arthroplasty **14**, 701–707 (1999)
17. Hamadouche, et al.: Alumina-on-alumina total hip arthroplasty: a minimum 18.5-year follow-up study. J. Bone Joint Surg. **84B**, 69–77 (2002)
18. Hannink, et al.: Transformation toughening in zirconia-containing ceramics. J. Am. Ceram. Soc. **83**(3), 461–487 (2000)
19. Huo, et al.: What's new in hip arthroplasty. J. Bone Joint Surg. Am. **88**, 2100 (2006)
20. Keurentjes, et al.: High incidence of squeaking in THAs with alumina ceramic-on-ceramic bearings. Clin. Orthop. Relat. Res. **466**, 1438–1443 (2008)
21. Kuntz, et al.: Current state of the art of the ceramic composite material BIOLOX®*delta*. In: Mendes, G, Lago, B. (eds.) Strength of Materials, Nova Science Publishers Inc., 133–155 (2009)
22. Kusaba, et al.: Abrieb von Hüftexplantat-Kugelköpfen aus Aluminiumoxidkeramik und Kobalt-Chrom im Vergleich. Deutscher Kongress für Orthopädie, Abstract, Berlin (2004)
23. Kusaba et al.: Alumina on alumina bearing with uncemented implant for dysplastic hips aged sixty or below: A five years minimum follow-up study to advantage the bearing property from a viewpoint of the surgeon. Abstract OSA04-03, ISTA 2008
24. Lewinnek, et al.: Dislocations after total hip-replacement arthroplasties. J. Bone Joint Surg. **60**, 217–220 (1978)
25. Lewis, et al.: Prospective randomized trial comparing alumina ceramic-on-ceramic with ceramic-on-conventional polyethylene bearings in total hip arthroplasty. J. Arthroplasty **25**(3), 392–397 (2010)
26. Lusty, et al.: Third-generation alumina-on-alumina ceramic bearings in cementless total hip arthroplasty. J. Bone Joint Surg. Am. **89**, 2676–2683 (2007)
27. Lusty, et al.: Minimising squeaking and edge loading when implanting a ceramic-on-ceramic hip arthroplasty. Touch Briefings, European Musculoskeletal Review. **1**, 73–75 (2007)
28. Magnissalis, et al.: Wear of retrieved ceramic THA components: four matched pairs retrieved after 5–13 years in service. J. Biomed. Mater. Res. **58**, 593–598 (2001)

29. Morlock, et al.: Mismatched wear couple zirconium oxide and aluminum oxide in total hip arthroplasty. J. Arthroplasty **16**, 1071–1074 (2001)
30. Murphy et al.: Clinical experience with the ceramic on ceramic articulation in THR in the USA. Abstract SA02–02, ISTA, Seoul, 2008
31. Nevelos, et al.: Analysis of retrieved alumina ceramic components from Mittelmeier total hip prostheses. Biomaterials **20**, 1833–1840 (1999)
32. Nizard, et al.: Ten-year survivorship of cemented ceramic-ceramic total hip prosthesis. Clin. Orthop. Relat. Res. **282**, 53–63 (1992)
33. Nizard, et al.: Alumina pairing in total hip replacement. J. Bone Joint Surg. Br. **87-B**(6), 755–758 (2005)
34. Olyslaegers, et al.: Wear in conventional and highly cross-linked polyethylene cups. J. Arthroplasty **23**(4), 489–494 (2008)
35. Oonishi, et al.: Alumina hip joints characterized by run-in wear and steady-state wear to 14 million cycles in hip-simulator model. J. Biomed. Mater. Res. **70**, 523–532 (2004)
36. Pandorf, T.: Wear of large ceramic bearings. In: Chang, J.D., Billau, K. (eds.) Bioceramics and Alternative Bearings in Joint Arthroplasty, Steinkopff-Verlag Darmstadt, 91–97 (2007)
37. Pokorny, et al.: The noisy hip – Is it only a ceramic issue? In: Cobb, J. (ed.) Modern Trends in THA Bearings. Material and Clinical Performance, Springer-Verlag Berlin Heidelberg, 85–90 (2010)
38. Prudhommeaux, et al.: Wear of alumina-on-alumina total hip arthroplasties at a mean 11-year follow up. Clin. Orthop. Relat. Res. **379**, 113–122 (2000)
39. Restrepo, et al.: The noisy ceramic hip: is component malpositioning the cause? J. Arthroplasty **23**(5), 643–649 (2008)
40. Restrepo, et al.: The effect of stem design on the prevalence of squeaking following ceramic-on-ceramic bearing total hip arthroplasty. J. Bone Joint Surg. Am. **92**, 550–557 (2010)
41. Rosneck, et al.: A rare complication of ceramic-on-ceramic bearings in total hip arthroplasty. J. Arthroplasty **23**(2), 311–313 (2008)
42. Sedel, et al.: Alumina–alumina hip replacement. J. Bone Joint Surg. Br. **72**, 639–658 (1990)
43. Sedel, L.: Evolution of alumina-on-alumina implants: a review. Clin. Orthop. Relat. Res. **379**, 48–54 (2000)
44. Simon, J., Dayan, A., Ergas, E., et al.: Catastrophic failure of the acetabular component in a ceramic-polyethylene bearing total hip arthroplasty. J. Arthroplasty **13**, 108 (1998)
45. Skinner: Ceramic bearing surfaces. Clin. Orthop. Relat. Res. **369**, 83–91 (1999)
46. Steppacher et al.: Outcome of ceramic-ceramic total hip arthroplastry at two to ten years in patients with developmental dysplasia of the hip. Abstract OSAA04-02, ISTA, Seoul, 2008
47. Toni, et al.: Early diagnosis of ceramic liner fracture. Guidelines based on a twelve-year clinical experience. J. Bone Joint Surg. Am. **88**, 55–63 (2006)
48. Toni, et al.: Osteolysis and ceramic-on-ceramic hip prostheses: "fact or fairy tale?". In: Cobb, J. (ed.) Modern Trends in THA Bearings. Material and Clinical Performance, Springer-Verlag Berlin Heidelberg, 55–60 (2010)
49. Walter, et al.: Edge loading in third generation alumina ceramic-on-ceramic bearings: stripe wear. J. Arthroplasty **19**, 402–413 (2004)
50. Walter, et al.: Squeaking in ceramic-on-ceramic hips: the importance of acetabular component orientation. J. Arthroplasty **22**, 496–503 (2007)
51. Widmer, KH.: "Safe Zone" in der HTEP: das zu erreichende Ziel. Leading Opinions **2**, 6–8 (2010)
52. Wollmerstedt, et al.: Aktivitätsmessung von Patienten mit Hüfttotalendoprothesen. Der Orthopäde **35**, 1237–1245 (2006)
53. Zichner, et al.: In-vivo-Verschleiß der Gleitpaarungen Keramik-Polyetyhlen gegen Metall-Polyethylen. [In vivo wear of ceramics-polyethylene in comparison with metal-polyethylene]. Der Orthopäde **26**, 129–134 (1997)
54. Fisher, et al.: Tribology of alternative bearings. Clin. Orthop. Relat. Res. **453**, 25–34 (2006)

55. Baldini et al.: Wear and tissue reaction in retrieved ceramic-on-ceramic THA. PF66, EFORT 2009
56. Capello, et al.: Ceramic-on-ceramic total hip arthroplasty:update. J. Arthroplasty **23**, 39–43 (2008)
57. Murphy, et al.: Two-to 9-year clinical results of alumina ceramic-on-ceramic THA. Clin. Orthop. Relat. Res. **453**, 97–102 (2006)
58. Mesko, et al.: Ceramic-on-ceramic hip outcome at a 5- to 10 year interval. Has lived up to its expectations? In: Cobb, J.P. (ed.) Modern Trends in THA Bearings Material and Clinical Performance, Springer-Verlag Berlin Heidelberg, 209–217 (2010)
59. Ihle, et al.: Keramik- und Metallköpfe im Dauertest – eine Langzeitanalyse des PE-Abriebs nach 20 Jahren. [Ceramic vs CoCrMo femoral heads in combination with Polyethylene cups. Long term wear analysis at 20 years.]. Orthopädische Praxis **46**(5), 221–230 (2009)
60. Descamps et al.: Comparative study of polyethylene wear in THR: 28 mm diameter ceramic versus metallic head:a fifteen years result. Abstract F67, EFORT, 2008
61. Dorlot, et al.: Wear analysis of retrieved alumina heads and sockets of hip prostheses. J Biomed Mater Res. **23** (A3 Suppl), 299–310 (1989)
62. Boehler, et al.: Long-term results of uncemented alumina acetabular implants. J. Bone Joint Surg. Br. **76**, 53–59 (1994)
63. Prudhommeaux, et al.: Wear of alumina-on-alumina total hip arthroplasties at a mean 11-year follow up. Clin. Orthop. Relat. Res. **379**, 113–122 (2000)
64. Hernigou, et al.: Ceramic-ceramic bearing decreases osteolysis: a 20-year study versus ceramic-polyethylene on the contralateral hip. Clin. Orthop. Relat. Res. **467**(9), 2274–2280 (2009)
65. Thomas, et al.: Orthopädisch-chirurgische Implantate und Allergien. Orthopade **1**, 1–14 (2008)
66. Raman et al.: Functional and clinical outcome of cementless primary arthroplasty of the hip using large diameter cermic bearing couples. Paper F244, EFORT 2010
67. Kim, Y.H., et al.: Cementless total hip arthroplasty with ceramic-on-ceramic bearing in patients younger than 45 years with femoral-head osteonecrosis. Int. Orthop. **34**(8), 1123–1127 (2010)
68. Kim, Y.H.: The results of a proximally-coated cementless femoral component in total hip replacement. J. Bone Joint Surg. Br. **90**, 299–305 (2008)
69. Urban, et al.: Ceramic-on-polyethylene bearing surfaces in total hip arthroplasty: Seventeen to twenty-one-year results. J. Bone Joint Surg. Am. **83**, 1688–1694 (2001)
70. Weber, et al.: Polyäthylen-Verschleiß und Spätlockerung der Totalprothese des Hüftgelenkes [Polyethylene wear and late loosening in total hip replacement]. Der Orthopäde **18**, 370–376 (1989)
71. Zagra et al.: Ceramic-ceramic coupling with big heads: clinical outcome. Eur J Orthop Surg Traumatol **17**(3), 247–251 (2007)
72. Dalla Pria, et al.: Evolution for diameter features and results. In: Chang, J.D., Billau, K. (eds.) Bioceramics and Alternative Bearings in Joint Arthroplasty, Steinkopff-Verlag Darmstadt, 99–105 (2007)
73. Zagra et al.: Ceramic-on-ceramic coupling with 36mm heads. P428, EFORT 2008
74. Giacometti-Ceroni, et al.: Alumina-on-alumina coupling with 36 mm heads. In: Zippel, H., Dietrich, M. (eds.) Bioceramics in Joint Arthroplasty, Steinkopff-Verlag, Darmstadt, 183 (2003)
75. Eskelinen, et al.: Total hip arthroplasty for primary osteoarthrosis in younger patients in the Finnish arthroplasty register. 4,661 primary replacements followed for 0-22 years. Acta Orthop. **76**, 28–41 (2005)
76. Schmalzried, et al.: The John Charnley Award. Wear is a function of use, not time. Clin. Orthop. Relat. Res. **381**, 36–46 (2000)
77. Schmalzried, T.P.: Quantitative assessment of walking activity after total hip or knee replacement. J. Bone Joint Surg. Am. **80**, 54–59 (1998)
78. The Swedish National Hip Register. Annual Report 2008

79. Clarke, et al.: Wear performance of 36 mm BIOLOX® forte/delta Hip combinations compared in simulated "severe" micro-separation test mode. In: Chang, J.D, Billau, K. (eds.) Bioceramics and Alternative Bearings in Joint Arthroplasty, Steinkopff-Verlag Darmstadt, 33–43 (2007)

80. Stewart, et al.: Long-term wear of ceramic-matrix composite material for hip prostheses under severe swing phase microsimulation. J. Biomed. Mater. Res. Appl. Biomater **66B**, 567–573 (2003)

81. Kinkel S.: Wollmerstedt N et al., Patient Activity after Total Hip Arthroplasty Declines with Advancing Age. Clin. Orthop. Relat. Res. published online, 27 February 2009

82. Wollmerstedt, et al.: The daily activity questionnaire. J. Arthroplasty **25**(3), 475–480 (2010)

83. Harris, W.H.: The problem is osteolysis. Clin. Orthop. Relat. Res. **311**, 46 (1995)

84. Schmalzried, et al.: Wear in total hip and knee replacements. J. Bone Joint Surg. Am. **81**, 115 (1999)

85. Kobayashi, et al.: Factors affecting aseptic failure of fixation after primary Charnley total hip arthroplasty. Multivariate survival analysis. J. Bone Joint Surg. Am. **79**, 1618 (1997)

86. Harris, W.H.: Wear and periprosthetic osteolysis. Clin. Orthop. Relat. Res. **393**, 66–70 (2001)

87. Phillips, et al.: Incidence rates of dislocation, pulmonary embolism, and deep infection during the first six months after elective total hip replacement. J. Bone Joint Surg. Am. **85**, 20–26 (2003)

88. Von Knoch, et al.: Late dislocation after total hip arthroplasty. J. Bone Joint Surg. Am. **84**, 1949–1953 (2002)

89. Mason, J.B.: The new demands by patients in the modern era of total joint arthroplasty: a point of view. Clin. Orthop. Relat. Res. **446**, 146–152 (2008)

90. Jasty, et al.: Wear of polyethylene acetabular components in total hip arthroplasty. An analysis of one hundred and twenty-eight components retrieved at autopsy or revision operations. J. Bone Joint Surg. Am. **79**(3), 349–358 (1997)

91. Kesteris, et al.: Polyethylene wear in Scanhip arthroplasty with a 22 or 32 mm head: 62 matched patients followed for 7–9 years. Acta Orthop. Scand. **67**(2), 125–127 (1996)

92. Eggli, et al.: Comparison of polyethylene wear with femoral heads of 22 mm and 32 mm. A prospective, randomized study. J. Bone Joint Surg. Br. **84**(3), 447–451 (2002)

93. Tarasevicius, et al.: Femoral head diameter affects the revision rate in total hip arthroplasty. An analysis of 1,720 hip replacements with 9–21 years of follow-up. Acta Orthop. Scand. **77**(5), 706–709 (2006)

94. Saikko, et al.: Low wear and friction in alumina/alumina total hip joint; a hip simulator study. Acta Orthop. Scand. **69**, 443–448 (1998)

95. Güttler, T.: Experience with BIOLOX®option revision heads. In: Benazzo, F, Falez, F, Dietrich, M. (eds.) Bioceramics and Alternative Bearings in Joint Arthroplasty, Steinkopff-Verlag, Darmstadt, 149–154 (2006)

96. Knahr, et al.: Revision strategies in total hip arthroplasty with respect to articulation materials. In: Benazzo, F, Falez F, Dietrich, M. (eds.) Bioceramics and Alternative Bearings in Joint Arthroplasty, Steinkopff-Verlag Darmstadt, 163–168 (2006)

97. Knahr et al.: Strategies for head and inlay exchange in revision hip arthroplasty. In: Chang, J.D, Billau, K. (eds.) Bioceramics and Alternative Bearings in Joint Arthroplasty, Steinkopff-Verlag Darmstadt, 275–280 (2007)

98. Walter, W.K.: Australian experience with ceramic systems. In: D'Antonio, J.A, Dietrich, M (eds.) Bioceramics and Alternative Bearings in Joint Arthroplasty, Steinkopff-Verlag Darmstadt, 113–115 (2005)

99. D'Antonio, et al.: Alumina ceramics bearings for total hip arthroplasty. Clin. Orthop. Relat. Res. **436**, 164–171 (2005)

Tribology of Metal-on-Metal Bearings

4

Jasper Daniel and Amir Kamali

4.1
History of Metal-on-Metal Bearings

The first total hip replacement (THR) was designed and implanted by Philip Wiles in the early 1930s, in Middlesex Hospital, London. The records of Wiles' cases were lost during the war but one patient was reported to still have the implant in situ 35 years later. Prior to Wiles' total hip replacements, only hemi-arthroplasty operations were performed, where one arthritic surface in the hip joint is replaced. The results of hemi-arthroplasty operations were unsatisfactory. McKee who had trained with Wiles began to develop his own THRs, including various uncemented devices in the 1940s and 1950s. However, these early prototypes only provided initial pain relief and they soon failed due to loosening and mechanical failure. Later iterations of the THR employed bone cement for fixation.

One such design was the McKee-Ferrar THR developed in the 1960s, which was the first widely used and successful THR. The McKee-Ferrar was a cobalt chrome molybdenum (CoCrMo) alloy, metal-on-metal (MoM) device with both femoral and acetabular components using cement fixation. However, the relatively simplistic manufacturing techniques used for MoM devices at the time led to poor clinical results. At the same time, developments and early successes in what is now termed 'conventional' metal-on-polyethylene (MoPE) hip replacements largely caused the discontinuation of MoM.

Young patients suffering with end stage osteoarthritis (OA) or rheumatoid arthritis (RA) of the hip require a solution that is bone conserving. Typically there is a loss of articular cartilage, a few millimeters thick, on the femoral and the acetabulum side. Hip resurfacing has been an attractive concept with many theoretical advantages which include minimal bone resection, restoration of normal femoral loading, prevention of stress shielding and restoration of normal

J. Daniel (✉)
Materials Science, University of Cambridge, Pembroke Street, CB2 3QZ, Cambridge
e-mail: jtd23@cam.ac.uk

A. Kamali
Implant Development Centre (IDC), Smith and Nephew Orthopaedics Ltd, Aurora House,
Spa Park, Harrison Way, Leamington Spa, CV31 3HL, UK
e-mail: amir.kamali@smith-nephew.com

K. Knahr (ed.), *Tribology in Total Hip Arthroplasty*,
DOI: 10.1007/978-3-642-19429-0_4, © 2011 EFORT

hip function with the added advantage of the option of revision to THR. The first resurfacing was performed by Sir John Charnley in the 1950s. He pioneered the double cup polytetrafluoroethylene (PTFE, also known as Teflon) device, due to the material's low coefficient of friction [9, 10]. However, the wear resistance of these components was low and they failed within 2 years in vivo. The 1970s saw the next major developments in artificial hip resurfacing technology, which employed materials which were already used in total hip replacements. Material combinations used included metal-on-polyethylene (MoPE) and ceramic-on-polyethylene components [1]. Freeman et al. used polyethylene femoral components against metal acetabular cups. However, the results were poor with a survival rate of 55% at 6–7 years [16]. The principle of damaged cartilage surface replacement remained popular till the 1980s however the wear related issues were difficult to overcome [2, 7, 16, 17].

Up until the 1980s due to the repeated failure of all the hip resurfacings, the concept was abandoned. It was believed that the hip resurfacing procedure would inevitably result in avascular necrosis (AVN) of the femoral head leading to femoral collapse. This theory was widely accepted and the concept of hip resurfacings looked dead in the water. However, later studies revealed the real cause for failure of MoPE resurfacings. It was discovered that the reason for the bone destruction and femoral collapse was osteolysis induced by wear particles [18, 20]. Later studies confounded the belief that all hip resurfacings would, in every case result in AVN, as the blood supply route for an arthritic femur is different to that of a normal healthy one. In a healthy femoral head, the blood supply is from extra-osseous vessels, traversing through the capsule and the synovium. In contrast, the majority of blood supply to an arthritic hip is intra-osseous, therefore enabling a surgical approach for a hip resurfacing arthroplasty without the risk of AVN of the femoral head [15].

In order for the concept of hip resurfacing to work, a bearing material had to be found, which could survive the rigours of high activity use for young patients whilst being sufficiently durable at large diameters. In order to avoid excessive resection of healthy bone, the material needs to have good mechanical properties even when thin walled. Metal-on-metal hip replacements have been used for more than four decades. The relatively simplistic early manufacturing techniques used and the poor clinical results than ensued, put an end to their use. However, there were instances when MoM implants were successful. Some components survived in vivo up to 30 years with minimal amount of volumetric wear. Three types of MoM devices, McKee-Farrar, Stanmore and Ring reported cases with no osteolysis and low wear, up to 37 years in vivo.

Lessons gleaned from the failures and successes of these first generation MoM devices were implemented in the development of the modern day MoM devices. The importance of manufacturing tolerances in terms of form, clearance, surface finish and component microstructure were studied. Retrospective studies were carried out in histology, engineering, materials and with experience in vivo, the Birmingham Hip resurfacing was launched in 1997 by Midland Medical Technology Ltd. The development of the Birmingham Hip resurfacing was based on a forensic study of first generation metal-on-metal hip replacement components: the Ring, the McKee-Ferrar and Stanmore hip replacement, which had been retrieved 20–37 years post implantation (Fig. 4.1). These components showed minimal wear rates. A 23.5 years retrieval showed 10 μm linear wear in the femoral component and 8 μm wear in the acetabular component, which is equivalent to 0.43 μm/year and 0.35 μm/year in the head and cup respectively.

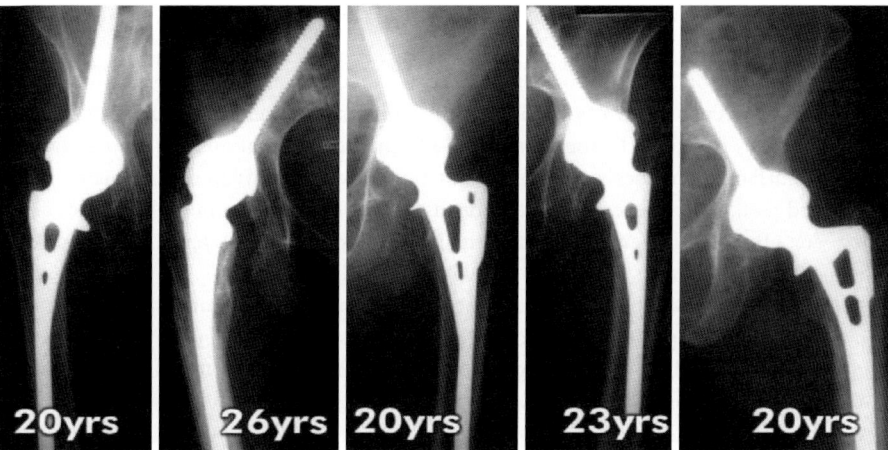

Fig. 4.1 Evidence of first generation large diameter metal-on-metal bearings showing little or no effects of metal wear over a long term in vivo

This device had a clearance of 272 μm, which is the difference between the diameter of the cup and the head (Fig. 4.8). These devices were made of high carbon grade (>0.2 wt.%) cobalt chromium molybdenum alloy. The devices are manufactured by the investment casting process, whereby a wax replica of the prosthesis is coated in a silica slurry. Once set, the wax in the slurry mold is heated to above its melting point (100–150°C), to remove it from the mold. This remaining mold is then filled with the molten metal alloy and allowed to cool. Once the metal is set, the slurry mold is then broken off. The cast metal is then put through a turning, honing, lapping and polishing process to produce the final product. The CoCrMo alloy used is biocompatible, resistant to wear and corrosion and has excellent mechanical properties for this application. The components have an as-cast (AC) microstructure and are not subjected to any thermal treatments. In this chapter, we will consider the main factors that influence the Tribological performance of MoM devices: wear, friction and lubrication.

4.2
Factors that Affect the Tribological Performance of MoM Devices

Surgeons and engineers developing the second generation MoM devices had the benefit of hindsight, to act on lessons learned about the various factors that influence wear, friction and lubrication. Design factors such as component clearance, articulation angle, radius, roundness and surface finish. Material specific factors which could affect tribology such as alloy composition and its microstructure. Most of the MoM devices on the market today differ slightly from each other in terms of design parameters which has led to differences observed in clinical outcomes.

4

4.2.1
Microstructure

Over the last decade there has been a lot of debate surrounding the subject of various microstructures of Cobalt Chromium Molybdenum (CoCrMo) alloys and their suitability for hip resurfacing prosthesis. The first generation of MoM total hip replacements, which performed well, were made of high carbon, as-cast cobalt chromium molybdenum alloy. There are several reports of those implants having lasted over 30 years in vivo with no evidence of osteolysis.

However, in an attempt to improve bearing performance, certain manufacturers of MoM devices developed different CoCrMo microstructures by means of varying the carbon content of the alloy or changing the processing methods e.g. thermal treatments to alter the metallurgy of as-cast CoCrMo.

Modern MoM devices are of high-carbon as-cast CoCrMo alloy (Fig. 4.2), which has a biphasic structure consisting of a face-centred cubic (FCC) austenitic matrix material which supports a second inter-dendritic metallurgical phase of chromium carbide (Fig. 4.3). The carbide phase has a coarse blocky morphology within the grains and at the grain boundaries. The carbides are mechanically stable and constitute approximately 5% of the bulk material in the as-cast state.

The first generation of MoM bearings which were clinically successful, were made of high-carbon as-cast (AC) CoCrMo alloy. However, in an attempt to optimise bearing design some manufacturers changed the microstructure of the alloy by varying the carbon content or the processing methods. Thermal treatments have also been used by some manufacturers in an attempt to reduce scrap resulting from porosity. The most common heat treatment process used is hot isostatic pressing followed by solution annealing. This process is called double heat treatment (DHT). For hot isostatic pressing (HIP) cast CoCrMo

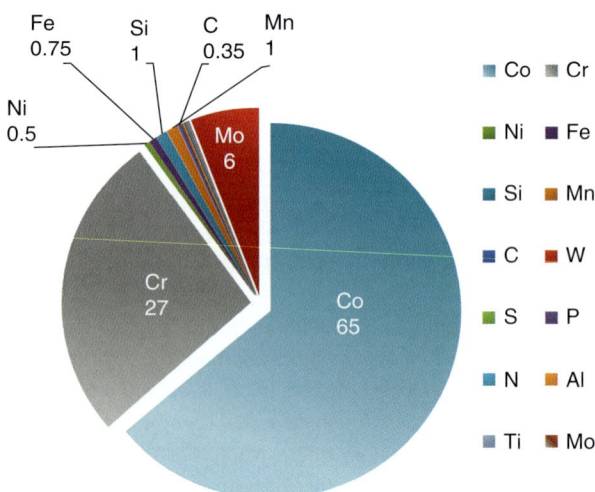

Fig. 4.2 Pie chart showing the percentages of constituent elements present in CoCrMo alloys used in MoM implants

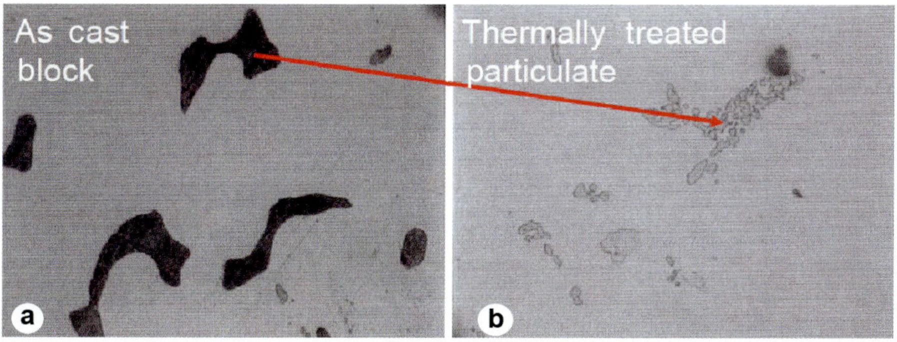

Fig. 4.3 Heat treatment reduces the size of the carbides and they become less mechanically stable and more prone to extraction. (**a**) as-cast microstructure and (**b**) double heat treated, CoCrMo microstructure

alloy is heated to 1200°C for 4 h at high pressure (103 MPa), in an inert atmosphere, followed by a gas fan quench at a slow cooling rate.

Pin-on-disc/plate and hip simulator studies have been carried out to determine the effect of the presence of carbides on wear of high carbon MoM components [8, 11, 28, 35, 44]. However, the effect of material microstructure on wear is apparent only when the components are not operating in complete fluid film lubrication (i.e. the two articulating surfaces are separated by a film of lubricant). A number of studies have reported no significant difference between the wear of different CoCrMo alloys' microstructures in hip simulator studies [5, 35]. However with MoM devices the effect of material microstructure is far less expressed in hip simulator studies due to the nature of fast, repeated and uninterrupted motions of the joint according to the ISO 14242–1 test protocol [23].

4.2.2
Articulation Angles

A well designed hip resurfacing component needs to maximise range of motion and articular coverage, in order to reduce the risk of edge loading and impingement or subluxation. Increasing the ÿ angle, would increase the risk of edge loading as the component would articulate closer to the edge of the cup (Fig. 4.4a). For optimum design, the acetabular component's articulation angle is a compromise between range of motion and risk of edge loading. The greater the articulation angle, lower the risk of edge loading; however this results in a reduced range of motion and increase risk of impingement. Excessive anteversion (>20°) and inclination (>45°) can lead to higher metal ion release in MoM bearings. Acetabular component position of 40–45° inclination and 15–20° anteversion is recommended for the Birmingham Hip Resurfacing (BHR) device.

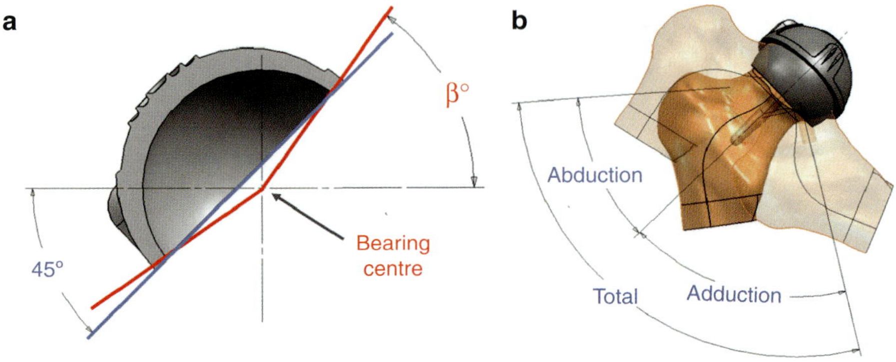

Fig. 4.4 It is important to achieve maximum range of motion, whilst providing the greatest amount of coverage to avoid edge loading. (**a**) The β angle illustrates the true inclination angle of an acetabular component from the bearing centre. (**b**) The range of motion is influenced by the head neck ratio and the acetabular component design

4.2.3
Surface Finish

Surface roughness plays a key part in determining the lubrication mode a hip replacement will operate under. Closer examination of the surface of a material will reveal features that can be characterised individually – these are called asperities. If the asperities present at the surface of the two articulating surfaces are larger than the fluid that separates them, then the wear and friction in the joint will be higher. Roughness values can be measured, by performing a linear trace along the surface of the material using a contacting probe (Fig 4.5).

It is a feature of MoM devices, under ideal conditions they operate predominantly under mixed lubrication regimes. This is the case when the bearings is in part separated by the fluid that is between articulating surfaces and in part supported by the asperities of the components. A complete fluid film can be observed in MoM components at high sliding velocities, high lubricant viscosity and at low loads.

MoM devices go through two phases of wear, the 'running-in' and 'steady' state wear. During the running in phase, the asperities on the surface of the metal surface are removed in a polishing process. The component wear rate is higher during the running in phase compared to the steady state phase. To this end, it is desirable to have a surface finish and texture that will most quickly transform the component from the 'running-in' phase to a 'steady' state phase with minimal amount of wear.

The following surface roughness parameters are of interest:

R_a, is the arithmetic average of absolute values (Fig. 4.6):

$$R_a = \frac{1}{n}\Sigma\mid y_i\mid \tag{1}$$

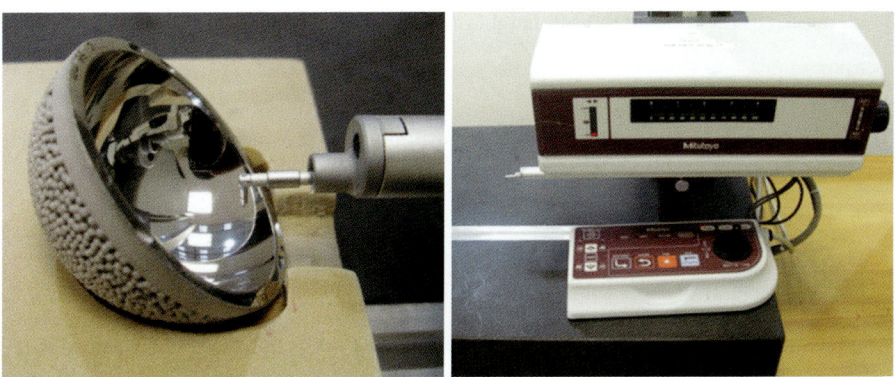

Fig. 4.5 A surface roughness measuring machine (Surftest SV3000). A 5 mm trace is taken across the surface of the component being measured. Values of R_a, R_{sk} and R_p are collected

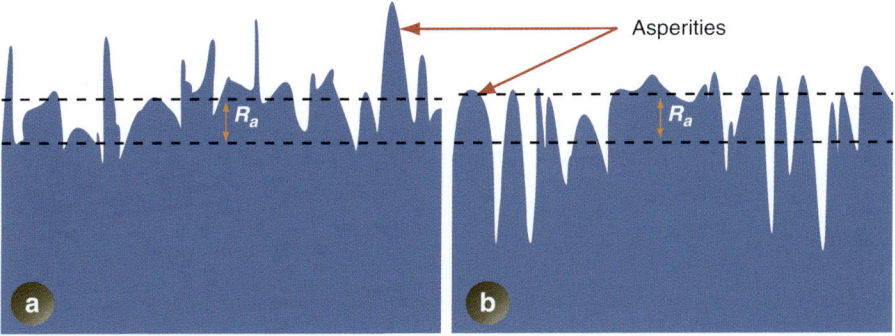

Fig. 4.6 Schematic of typical surface profiles of metal-on-metal components. Surface roughness, R_a which is the arithmetic average of absolute values. R_{sk} describes the skew of the values (**a**) positively skewed and (**b**) negatively skewed

R_a does not always give the complete picture, Fig. 4.6 shows how two different profiles (a and b) can result in the same R_a value. Therefore it is useful to get an idea of the skewness of the surface that is being measured.

R_{sk} describes the skew of the values

$$R_{sk} = \frac{1}{nR_q^3} \sum y_i^3 \qquad (2)$$

R_{sk} can be defined as the measure of the symmetry of a profile about a mean line. It can distinguish between asymmetrical profiles which have the same R_a values. So if the profile is predominantly valleys, it is negative skew and positive skew is mainly peaks. Examples of positive and negative skew are given in Fig. 4.6a and b respectively.

R_p describes the maximum peak height of the measured profile.

$$R_p = \max_i y_i \tag{3}$$

These three parameters used together form a useful image of what the surface of the material looks like.

4.2.4
Roundness

Along with surface finish, another important factor that needs to be tightly controlled is the roundness of the component. A profile is generated by a roundness machine, where a probe measures the roundness of the device which is placed on an air bearing spindle (Fig. 4.7). An average roundness value is obtained by gathering a large number of measured points from the bearings surface which are then used to calculate a best-fit circle through them. Also the maximum deviation of individual points from this best-fit circle is calculated (RONT). When the form value is correctly specified and small enough, it ensures that misshaped components can easily be identified. Errors in the shape can results in the device clearance no longer being constant for all contact positions of the bearing. In MoM bearings where the clearance is very tight, it is important that the roundness is as near perfect as possible.

4.2.5
Clearance

Modern MoM bearings conform to a polar bearing configuration wherein curvature of the cup is larger than that of the head. The other two configurations (i.e. equatorial and annular)

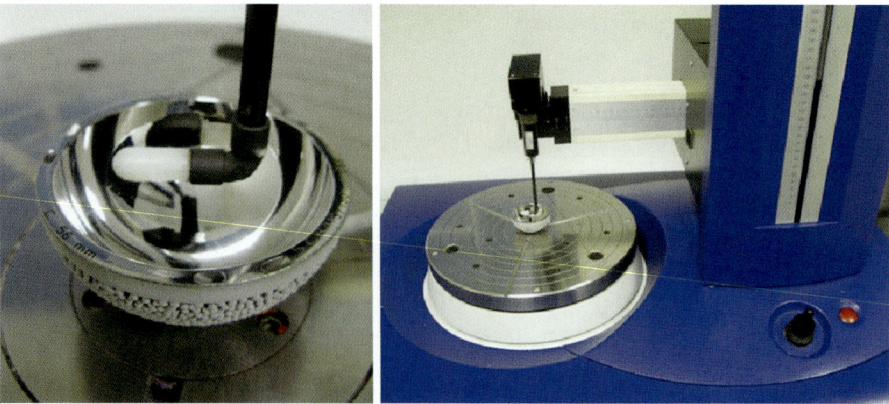

Fig. 4.7 Roundness measuring equipment. The component is held securely on an air table and a profile is taken to ascertain the deviation of the component from a perfect circle. The machine has an accuracy of 0.1 μm

Fig. 4.8 Diametral clearance is twice the radial clearance and is defined as the difference between the inner diameter of the cup and the diameter of the head i.e., $2(R_2 - R_1)$

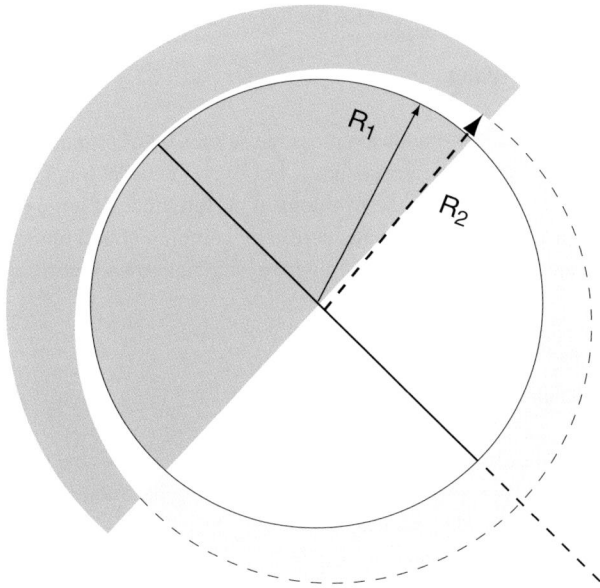

generate high friction and wear and are therefore not suitable for MoM devices. Clearance is defined as the difference between the diameter of the cup and the head of the device and affects the lubrication (Fig. 4.8). Clearance is one of the most influential characteristics of a MoM device. However, there has been a lot of debate about what is the ideal clearance range for MoM devices [6, 14, 33, 42].

Hip simulator studies have consistently shown a significant reduction in wear when smaller clearances are employed. Lower clearance components in the presence of joint lubricant can generate better lubrication conditions and therefore protect the articulating surfaces from friction and wear. Traditionally the lubricant used in hip simulators to study friction employ bovine serum (in some cases with carboxy methyl cellulose), as this is representative of the viscosities of human synovial fluid. Viscosity is an important parameter in determining the thickness of the fluid film generated and therefore the friction of the bearing. Soon after joint replacement surgery, the implant is bathed in blood and not synovial fluid. Blood is a complex fluid which contains macromolecules and various white and red blood cells up to 20 μm in size. MoM cups are press fitted into an under-reamed acetabulum, and are reliant on a good primary fixation in order for the component to be stabilised in the early post operative period. A study was devised to determine the optimum clearance of a MoM devices, when it is in this crucial early post operative period, when blood is the lubricant. This study showed that lower clearance devices had greater friction when using whole blood as lubricant [33].

In this section we have seen how design and processing factors have an impact on the performance of MoM devices. In the next part of this chapter, we will see how our understanding of these parameters has been expanded through in vitro testing of MoM devices.

4

4.3
In Vitro Studies

Basic testing methods to assess the suitability of implant materials in vitro have existed as long as hip replacements have. For over a decade, hip simulators of various designs have been used to assess the tribological performance of large diameter MoM devices. The use of hip simulators, in their increasing complexity and physiological relevance, has led to a greater understanding of implant wear, friction and lubrication.

4.3.1
Friction and Lubrication

Friction can be defined as the tangential resistance as a result of relative motion or motion between two surfaces. The laws of friction, between two dry surfaces were established by Amontons and Coulomb, where
   Friction (F) is:

1. Directly proportional to the normal load applied (W);
2. Independent of the apparent area of contact;
3. Independent of sliding speed (V)

   Therefore,

$$F = \mu W \tag{4}$$

Where μ is known as the coefficient of friction. There are several factors that contribute to frictional force; molecular interaction between surfaces, adhesion, deformation and the presence of surface asperities at the molecular level of a seemingly smooth surface. The friction coefficient in MoM devices is influenced by various factors including load applied, material hardness, surface characteristics, bearing diameter, sliding speed of the two counter surfaces, radial clearance and the lubricating properties of joint fluid. Some of these properties are patient dependent and therefore cannot be controlled, others properties are within the control of engineers such as material, radial clearance and surface characteristics. When frictional torque is sufficiently large, micromotion can occur and it can lead to mechanical loosening of the implant [38].

   Under non lubricated conditions, friction and wear are directly proportional to a product of the load applied and the diameter of the bearing. Therefore, for a given load, larger diameter bearings should generate more wear than smaller diameter bearings. However, in the presence of a joint lubricant, that rule is not sustained.

   Under those conditions, the wear and friction are governed by the thickness of the fluid film that can be generated and in particular on a ratio (λ) of the minimum film thickness (h_{min}) possible and the average surface roughness (R_a) of the material. If, $\lambda < 1$, the surfaces are in sliding contact with each other and boundary lubrication is inevitable. If $\lambda > 3$, then the surfaces are separated from each other, providing fluid film protection of the two articulating surfaces. If $3 > \lambda > 1$, then the bearing is operating under a mixed lubrication regime.

The equation that governs h_{min} is as follows:

$$\frac{h_{min}}{R} = 2.8 \left(\frac{n\mu}{E'R}\right)^{0.65} \left(\frac{w}{E'R^2}\right)^{-0.21} \tag{5}$$

Where

$$R_x = \frac{R_1 R_2}{R_1 - R_2} \tag{6}$$

And

$$\frac{1}{E'} = \frac{1}{2}\left[\frac{1-\sigma_1^2}{E_1}\right]\left[\frac{1-\sigma_2^2}{E_2}\right] \tag{7}$$

R_1 and R_2 are the radii of curvature, E_1 and E_2 are the Young's moduli and σ_1 and σ_2 are the Poisson's ratio of the cup and the head respectively, h is the viscosity of the fluid and L is the load applied. From the equation, it can be deduced that, all other factors being equal, the smaller the radial clearance the greater the potential to generate a fluid film and therefore the lower the wear and friction.

Minimizing friction during the early post operative period allows for bony in-growth on to the back surface of the cup, which provides long term fixation. Another consequence of interference press fitting the cup in to the acetabulum, is the deformation of the implanted cup. The anatomy of the pelvis results in non uniform loading of the acetabulum cup. Intra-operative measurements have shown up to 100 μm of cup deformation immediately after implantation of the cup. The Birmingham hip resurfacing is a cementless, interference press-fitted cup which relies on a good primary fixation in order for the cup to be stabilised.

In a separate clinicoradiologic study, low clearance components (100 μm) were implanted and patients followed up to assess in vivo wear performance in terms of metal ions [32]. However, at follow up, some patient radiographs showed a progressive radiolucent line in zone 1 and 2 of the acetabulum. These features were not previously observed with the standard BHR components and may be the result of increased friction in the early post operative period.

A study which considered the effect of clearance, joint fluid and the effect of deformation showed that the addition of hyaluronic acid (which contains macromolecules) to bovine serum leads to an increase in friction. Furthermore, when the same test was carried out using whole blood and clotted blood, the friction of the lower clearance component increased significantly. When whole blood was used as lubricant, a trend of reduction in friction coefficient was observed with an increase in clearance [33].

Cup deformation had a greater impact on the friction of low clearance components (Fig. 4.9). Whereas, the effect of deformation on the friction of larger clearance components in clotted blood was not noticeable.

4

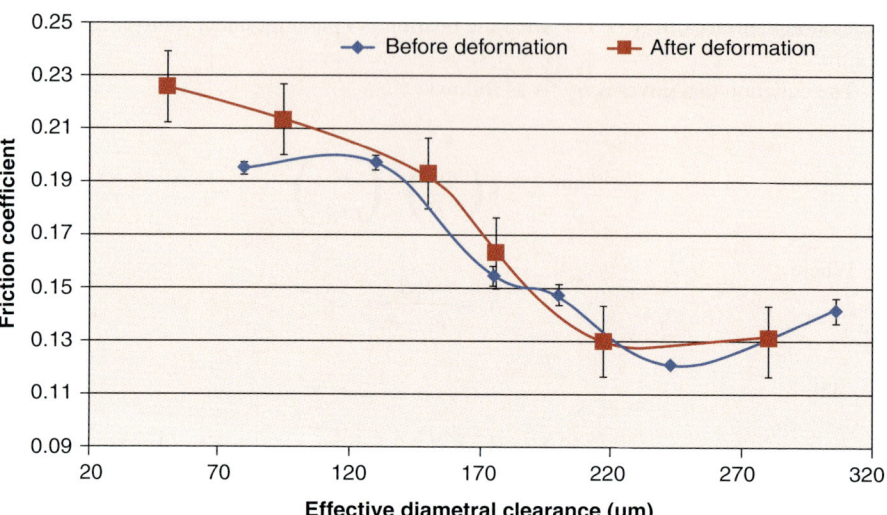

Fig. 4.9 Friction coefficient versus joint clearance for standard BHR cups and after 25–35 μm deformation using clotted blood as lubricant [26]

It has been reported that reduced clearance results in reduced wear [14, 43]. However, factors such as cellular and macromolecular shear and the effect of cup deformation which may affect component friction have not been considered in the past. This study showed that reduced clearance components have the potential to generate higher friction especially in the early weeks after implantation when the cup is deformed and in the presence of clotted blood (and not synovial fluid) as lubricant [24]. This increase in friction results in increased frictional torque at the bone-implant interface.

4.3.2
Wear

Tribological testing has been carried out for decades in an attempt to predict the longevity of various designs and materials in hip arthroplasty. Hip wear simulators have been used extensively as an advanced tool to determine the wear of implants under conditions that are considered to be close to that of normal physiological walking cycles (Fig. 4.10). In vitro wear studies however have consistently reported wear rates that are lower than those reported in clinical studies.

The accelerated wear studies carried out in hip simulators use uninterrupted and identical motions every cycle. This results in the joint operating predominantly under exaggerated lubrication conditions, which protects the bearing surfaces from wear. This is in contrast to the extensive range of motion (which includes stop start motion) and high force experienced in vivo, during an average day, resulting in the breakdown of the fluid film lubrication in MoM bearings. These less favourable lubrication conditions ultimately generate greater friction and wear in the components in vivo compared to that of the predicted values in vitro. In

Fig. 4.10 Hip simulator test setup (**a**) anatomical loading of the cup and the head (**b**) the femoral head fixture (**c**) the acetabular cup fixture and (**d**) the hip simulator in operation; the device is contained in a bag of simulation joint fluid of bovine serum

order to improve the predictive accuracy of hip simulator testing, a more physiologically relevant test protocol was developed [26] for MoM bearings, based on the study of patient activity at various stages of follow up [12].

One of the novel aspects of the test protocol was to employ a test frequency of 0.5 Hz (30 hip cycles/min) with stop/start motion added every 100 cycles [12, 41]. Two sets of kinematics and kinetics were used side by side, and they were alternated every 100 cycles in order emulate the variations in patients' day to day activity. The first, a Paul-type stance phase loading was used, with maximum load applied was 3 kN and the minimum load during the swing phase was 0.3 kN [45]. The flexion/extension range used was 30°/15° with an internal external rotation of ±10°. In the second set of kinetics and kinematics, flexion/extension range was ±22° and internal/external rotation was ±8° [22, 34], maximum stance phase load of 2.2 kN and a swing phase load of 0.24 kN [4].

In this study, 50 mm MoM devices were tested, three as-cast (AC) devices and four double heat treated (DHT). The AC and DHT components used had similar clearances, mean clearance 234 μm (±5.7 μm) and 241 μm (±1.0 μm) respectively. The test lubricant collected and replenished with new lubricant every 125 k cycles. The test was taken down and the components cleaned and gravimetric measurements were taken at 0.5, 1.0, 1.5 and 2 million cycles. Gravimetric measurements were taken following the ISO 14242–20 protocol [21]. The test lubricant that was collected and analysed for cobalt, chromium and molybdenum ion levels using a high resolution inductively coupled plasma mass spectrometer (HR-ICPMS, Element Thermo Finnigan MAT, Bremen, Germany).

The wear results of the hip simulator study show a typical biphasic wear pattern, this is similar to the pattern observed in vivo. The two phases of wear have been described as

4

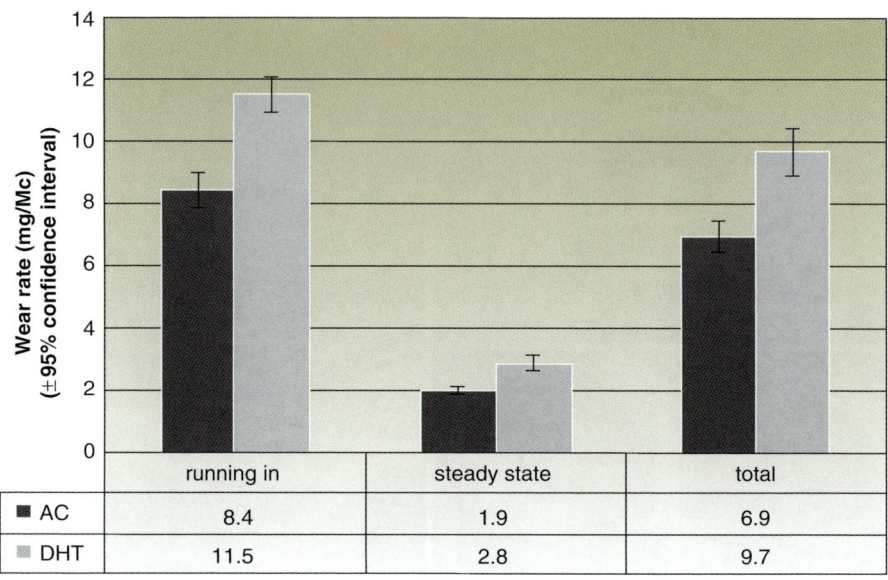

	running in	steady state	total
■ AC	8.4	1.9	6.9
▨ DHT	11.5	2.8	9.7

Fig. 4.11 Gravimetric wear rate of the as-cast and double heat treated components. The running in phase and the steady state phase are defined at 0–0.5 Mc and 0.5–2 Mc, respectively [26]

'running-in' (0–0.5 Mc) and a 'steady-state' (0.5–2 Mc) phase. The mean gravimetric wear rates during the running-in phase of the DHT and AC devices were calculated as 11.5 mg/Mc and 8.4 mg/Mc respectively (Fig. 4.11). The difference in running-in wear for the AC and DHT component groups was statistically significant ($p=0.014$).

The steady state wear of the AC components, 1.9 mg/Mc, was statistically less than ($p=0.002$) that of the DHT components, 2.8 mg/Mc. The metal ions results were calculated using the combined ion levels of cobalt, chromium and molybdenum ions measured in the test lubricant. The metal ions results showed a similar biphasic wear trend as the gravimetric wear result did. However, even though the trends were similar, the measured metal ions levels were lower than the gravimetric wear results. This may be due to the incomplete ionisation of all the particles in the lubricant during the measuring process. The mean wear rate during the running in phase for the AC devices was measured at 6.3 mg/Mc and 8.2 mg/Mc for the DHT devices (Table 4.1). However, the different wear rates for the two microstructures during the running in phase is not statistically significant ($p=0.159$). The steady state metal ion levels for the AC and the DHT components was 0.6 and 1.8 mg/Mc respectively. The metal ion levels of the DHT devices was three times greater than the AC components and this difference during the steady state phase was statistically significant ($p=0.024$). The increase in metal ions generated by the DHT devices compared to the AC components may be explained by a previous study which showed DHT devices generated smaller particles and in much greater numbers [28]. This would result in a greater surface area of metal particles which are susceptible to corrosion and thereby increase in metal ions.

A recent study reported the clinical outcomes of operations performed by a single surgeon comparing AC and DHT devices [13]. The study reported up to 24% of the patients, who received the DHT components, showed radiological signs of failure. The results of

Table 4.1 Mean rate of release of metal ions using the as-cast and the double heat-treated devices during the running-in and steady-state phases [26]

Devices	Mean metal ion rate (mg/Mc) (±95% confidence interval)	
	Running-in phase	Steady-state phase
As-cast ($n=3$)	6.3±(1.9)	0.6±(0.1)
Double heat treated ($n=4$)	8.2±(1.4)	1.8±(0.6)

this hip simulator may explain the higher incidence of osteolysis and component loosening observed in DHT components.

4.4
Implant Orientation

Retrieval studies have shown that implant orientation, particularly inclination angle of the acetabular cup is essential for minimizing wear and risk of dislocation. Correct implant orientation also provides maximum range of motion and increases the longevity of the implant. A hip simulator study was carried out to determine the effect of cup orientation on the wear of the MoM devices [25]. Ten, 50 mm BHR devices were tested, divided into three groups ($n=3$/group) of varied orientations and with one control device (Fig. 4.10). For the three groups, the distances between the edge of the cup and the wear patch were varied and its impact on wear was assessed.

All the devices showed a bi-phasic wear pattern with a relatively higher wear rate at the at start of the test called the 'running-in' phase, followed by a lower linear wear rate called the 'steady-state' phase. The joints showed no significant difference between the groups during their running in period, nor were there any significant differences during the steady state period of the test between the three groups ($p>0.05$). When articulation is limited to being within the bearing surfaces of the prosthesis, changes in distance of articulation from the edge of the cup had no impact on wear in this study. However when the researchers introduced impingement, it resulted in edge loading (head articulating on the edge of the cup) for one of the devices. The edge loaded component had a 60-fold increase in wear as a result of edge loading. It is important to note that high wear may result in high metal ions and adverse tissue reactions.

4.5
In Vivo Studies

4.5.1
Patient Activity Levels at Follow up

In a study carried out by the McMinn Centre, Birmingham, UK, patient activity levels were assessed for a group of 28 preoperative patients and 183 patients with a unilateral BHR arthroplasty at different stages of follow-up between 1–10 years [12]. Mean age of

4

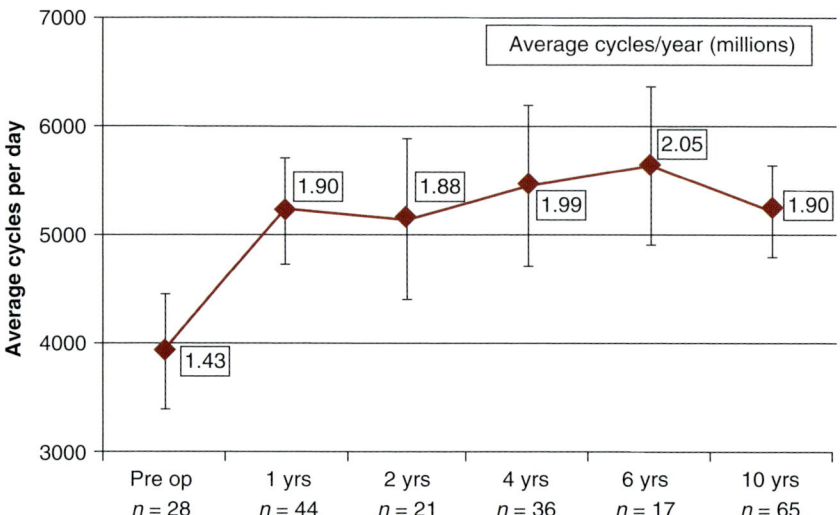

Fig. 4.12 A cross-sectional study of the activity levels of 28 preopeative and 183 unilateral BHR patients at various stages of follow-up. Mean daily step activity rates ±95% confidence intervals [12]

the patients at the time of follow-up was 54.4 years (range 32–65). The patients were advised to wear a step activity monitor (SAM, Cymatech. Seattle WA, USA) just above the lateral malleolus of the right leg or the medial malleolus of the left leg over a period of 5–7 days throughout the waking hours of the day. The temporal trend in change of step activity in these patients was noted. The aim of this study was to assess if the activity levels of patients with Birmingham Hip Resurfacing (BHR) arthroplasties are sustained as the follow-up progresses.

This study provided objective evidence of the activity rates of patients at different stages of follow-up after a MoM surface replacement arthroplasty. The findings of this study demonstrate a significant improvement in patient activity following the operation (Fig. 4.12). Over the subsequent 10-year period there is no significant reduction of activity levels in this patient group who are 65 years or younger. In spite of the excellent patient activity follow up data and clinical survivorship of these devices, in some rare cases MoM implants are known to initiate adverse tissue reactions.

4.5.2
Adverse Tissue Reactions and Pseudotumours

Metal-on-metal wear leads to the release of metal particles which are insoluble and metal ions which are soluble. Metal ions are responsible for systemic exposure as they are able to enter the systemic circulation of the body. However there is an effective renal clearance mechanism for these ions, hence avoiding cumulative build up of metal ions in the body.

Adverse tissue reactions and pseudotumours are not a new phenomenon in relation to hip replacement arthroplasty. Conventional MoPE hip and knee bearings have been the reason for pseudotumours for several years [19, 39]. More recently cases of pseudotumours have been reported relating to patients with metal-on-metal hip prosthesis [30, 36]. Pseudotumours are usually granulomatous lesions that develop in the peri-prosthetic region and resemble in appearance a tumour. They can be varied in size and constitution and may or may not be in contact with the prosthesis.

Studies have shown that pseudotumours in metal-on-metal hip resurfacing arthroplasty are associated with high levels of metal ions measured in vivo, thus implying that they occur in high wear components. In a recent study that aimed to quantify the wear of components that were revised due to pseudotumours compared to controls components that were revised for other reasons. The components revised due to pseudotumours were found to have significantly higher rates of wear compared to the control group. All the retrieved implants with incidence of pseudotumours, had been edge loaded [29]. Therefore, edge-loading which results in undesired lubrication conditions may be the cause for high wear and consequently in some cases leads to pseudotumours.

4.5.3
Clinical Outcomes of MoM

There are many different designs of MoM bearings with various metallurgies as previously discussed. These differences have resulted in variations in clinical performance of these devices. Reviewing the published clinical data, it is clear that the Birmingham hip resurfacing has the most successful performance of the MoM devices (No authors listed, 2010; [37]). Since the first BHR was implanted in July of 1997, more than 130,000 operations have been performed worldwide. The BHR was the first MoM hip resurfacing to receive food and drug administration (FDA) approval for use in the United States, in May 2006. Studies from all around the world have shown the BHR performing consistently well [27, 31, 40]. The Australian joint registry in 2009 reported on 8,427 hips replaced, a 95% survivorship at 8 years for the BHR, which out performed all its competitors [3]. The Oswestry registry, reported 95.4% survivorship at a minimum of 10 years collected from 17 countries and 21 different surgeons on the same device [40]. Other MoM devices have not shared the same degree of success – owing to differences in design aspects such as articulation angle, clearance and materials microstructure variations. BHR hip resurfacing has continued to perform at a high standard across the globe. It is increasingly being adopted and has revolutionised the way young and active hip arthritis patients are treated.

References

1. Amstutz, H.: Surface replacement arthroplasty. In: Amstutz, H.C. (ed.) Total Hip Arthroplasty, pp. 295–332. Churchill Livingstone, Edinburgh (1991)
2. Amstutz, H.C., Graffradford, A., Gruen, T.A., Clarke, I.C.: Tharies surface replacements – review of First 100 cases. Clin. Orthop. Relat. Res. **134**, 87–101 (1978)

3. No authors listed.: Australian Orthopaedic Association National Joint Replacement Registry: Hip and Knee Arthroplasty Annual Report 2010, URL http://www.aoa.org.au/jointregistry-pub.asp (2010)
4. Bergmann, G., Graichen, F., Rohlmann, A.: Hip joint loading during walking and running measured in two patients. J. Biomech. **26**, 969–990 (1993)
5. Bowsher, J.G., Williams, P.A., Clarke, I.C., Green, D.D., Donaldson, T.K.: "Severe" wear challenge to 36 mm mechanically enhanced highly crosslinked polyethylene hip liners. J. Biomed. Mater. Res. B Appl. Biomater. **86**, 253–263 (2008)
6. Brockett, C.L., Harper, P., Williams, S., Isaac, G.H., Dwyer-Joyce, R.S., Jin, Z., Fisher, J.: The influence of clearance on friction, lubrication and squeaking in large diameter metal-on-metal hip replacements. J. Mater. Sci. Mater. Med. **19**, 1575–1579 (2008)
7. Capello, W., Misamore, G., Trancik, T.: Conservative total hip-arthroplasty. Orthop. Clin. North Am. **13**, 833–842 (1982)
8. Cawley, J., Metcalf, J.E.P., Jones, A.H., Band, T.J., Skupien, D.S.: A tribological study of cobalt chromium molybdenum alloys used in metal-on-metal resurfacing hip arthroplasty. Wear **255**, 999–1006 (2003)
9. Charnley, J.: Arthroplasty of the hip. a new operation. Lancet **280**, 1129–1132 (1961)
10. Charnley, J.: Tissue reactions to polytetrauorethylene. Lancet **285**, 1379 (1963)
11. Clemow, A.J.T.: The influence of microstructure on the adhesive wear resistance of a co-cr-mo alloy. Wear **61**(2), 219–231 (1980)
12. Daniel, J., Kamali, A., Ziaee, H., McMinn, D.: Step activity monitoring of hip resurfacing patients at different stages following operation. Orthop. Res. Soc. Transactions vol. 34, Las Vegas, NV, 0366 (2009)
13. Daniel, J., Ziaee, H., Kamali, A., Pradhan, C., Band, T., McMinn, D.: Ten-year results of a double heat- treated metal-on-metal hip resurfacing. J. Bone Joint Surg. Br. **92**(1), 20–27 (2010)
14. Farrar, R., Schmidt, M.: The effect of diametral clearance on wear between head and cup for metal-on-metal articulations. Trans. 42nd Annual Meeting, Orthop. Res. Soc. 71 (1997)
15. Freeman, M.: Some anatomical and mechanical considerations relevant to surface replacement of femoral-head. Clin. Orthop. Relat. Res. (**134**, 19–24 (1978)
16. Freeman, M., Cameron, H., Brown, G.: Cemented double cup arthroplasty of hip – 5 year experience with iclh prosthesis. Clin. Orthop. Relat. Res. **134**, 45–52 (1978)
17. Furuya, K., Tsuchiya, M., Kawachi, S.: Socket cup – arthroplasty. Clin. Orthop. Relat. Res. **134**, 41–44 (1978)
18. Goldring, S., Schiller, A., Roelke, M., Rourke, C., O'Neil, D., Harris, W.: The synovial-like membrane at the bone-cement interface in loose total hip replacements and its proposed role in bone lysis. J. Bone Joint Surg. Am. **65**, 575–584 (1983)
19. Griffiths, H., Burke, J., Bonfiglio, T.: Granulomatous pseudotumors in total joint replacement. Skeletal Radiol. **16**(2), 146–152 (1987)
20. Howie, D.W., Cornish, B.L., Vernon-Roberts, B.: Resurfacing hip arthroplasty. Classification of loosening and the role of prosthesis wear particles. Clin. Orthop. Relat. Res. **255**, 144–159 (1990)
21. ISO ISO 14242–2:2000: Implants for surgery: wear of total hip-joint prostheses. Part 2: methods of measurement. URL http://www.iso.org Date last accessed 3 March 2010
22. Johnston, R., Smidt, G.: Measurement of hip-joint motion during walking – evaluation of an electrogoniometric method. J. Bone Joint Surg. Am. **51**(6), 1082–1094 (1969)
23. Kaddick, C., Wimmer, M.A.: Hip simulator wear testing according to the newly introduced standard iso 14242. Proc. Inst. Mech. Eng. H **215**, 429–442 (2001)
24. Kamali, A., Daniel, J.T., Youseffi, M., Javed, S., Hussain, A., Chenxi, L., Band, T., Ashton, R., McMinn, D.: Cup deformation and its effect on friction in metal-on-metal bearings. 55th Annual Meeting. Orthop. Res. Soc., Poster No. 2219 (2009)

25. Kamali, A., Hussain, A., Li, C., Ashton, R., Band, T.: The effect of cup orientation on the Tribological performance of 50 mm Birmingham hip resurfacing (BHR). 54th Annual meeting. Orthop. Res. Soc. (2008)
26. Kamali, A., Hussain, A., Li, C., Pamu, J., Daniel, J., Ziaee, H., Daniel, J., McMinn, D.: Tribological performance of various cocr microstructures in metal-on-metal bearings. J. Bone Joint Surg. Br. **92**, 717–725 (2010)
27. Khan, M., Kuiper, J., Edwards, D., Robinson, E., Richardson, J.: Birmingham hip arthroplasty: five to eight years of prospective multicenter results. J. Arthroplasty **24**, 1044–1050 (2009)
28. Kinbrum, A., Unsworth, A.: The wear of high-carbon metal-on-metal bearings after different heat treatments. Proc. Inst. Mech. Eng. H **223**(1), 131 (2009)
29. Kwon, Y., Glyn-Jones, S., Simpson, D., Kamali, A., Counsell, L., Mclardy-Smith, P., Beard, D., Gill, H.: Analysis of wear of retrieved metal-on-metal hip resurfacing implants revised due to pseudotumors. J. Bone Joint Surg. Br. **92**(3), 356–361 (2010)
30. Murray, D.: Analysis of wear of retrieved metal-on-metal hip resurfacing implants revised due to pseudotumors. J. Bone Joint Surg. Br. **92**(3), 356–361 (2010)
31. McBryde, C., Theivendran, K., Thomas, A., Treacy, R., Pynsent, P.: The influence of head size and sex on the outcome of Birmingham hip resurfacing. J. Bone Joint Surg. Am. **92**, 105–112 (2010)
32. McMinn, D., Daniel, J.: History and modern concepts in surface replacement. Proc. Inst. Mech. Eng. H **220**(2), 239–251 (2006)
33. McMinn, D., Daniel, J., Kamali, A., Saravi, S.S., Youseffi, M., Daniel, J., Band, T., Ashton, R.: Friction testing in metal-metal bearings with different clearances using blood as lubricant. 52nd Annual Meeting 90. Orthop. Res. Soc. 31 (2006)
34. Murray, M.: Gait as a total pattern of movement. Am. J. Phys. Med. **46**, 290–333 (1967)
35. Nevelos, J., Shelton, J.C., Fisher, J.: Metallurgical considerations in the wear of metal-on-metal hip bearings. Hip Int. **14**, 1–10 (2004)
36. Pandit, H., Glyn-Jones, S., McLardy-Smith, P., Gundle, R., Whitwell, D., Gibbons, C., Ostlere, S., Athanasou, N., Gill, H., Murray, D.: Pseudotumours associated with metal-on-metal hip resurfacings. J. Bone Joint Surg. Br. **90**(9), 847–851 (2008)
37. Porter, M., Borroff, M., Gregg, P., Howard, P., MacGregor, A., Tucker, K.: National Joint Registry for England and Wales: 7th Annual Report. (2010)
38. Resnick, B., Nahm, E.S., Orwig, D., Zimmerman, S.S., Magaziner, J.: Measurement of activity in older adults: reliability and validity of the step activity monitor. J. Nurs. Meas. **9**(3), 275–290 (2001)
39. Ritter, M.: The anatomical graduated component total knee replacement: a long-term evaluation with 20 year survival analysis. J. Bone Joint Surg. Br. **91**, 745–749 (2009)
40. Robinson, E., Richardson, J., Khan, M.: Minimum 10 year Outcome of Birmingham Hip Resurfacing (BHR), A review of 518 Cases from an International Register. Oswestry Outcome Centre, Oswestry (2009)
41. Silva, M., Shepherd, E., Jackson, W., Dela Rosa, M., Schmalzried, T.: Wear is a function of the amount and pattern of activity: The Step Activity Monitor (SAM). Proceedings of the 47th Annual Meeting, Orthop. Res. Soc. (2001)
42. Tuke, M.A., Scott, G., Roques, A., Hu, X.Q., Taylor, A.: Design considerations and life prediction of metal-on-metal bearings: the effect of clearance. J. Bone Joint Surg. Am. **90**, 134–141 (2008)
43. Unsworth, A., Pearcy, M., White, E., White, G.: Frictional properties of artificial hip joints. Eng. Med. **17**(3), 101–104 (1988)
44. Varano, R., Bobyn, J., Medley, J., Yue, S.: Does alloy heat treatment influence metal-on-metal wear? Orthop. Res. Soc. New Orleans, 28 (2003)
45. Williams, S., Butterfield, M., Stewart, T., Ingham, E., Stone, M., Fisher, J.: Wear and deformation of ceramic-on-polyethylene total hip replacements with joint laxity and swing phase microseparation. Proc. Inst. Mech. Eng. H **217**(2), 147–153 (2003)

Highly Cross-Linked Polyethylenes

5

Robert M. Streicher

5.1
Introduction

Several issues continue to affect the survival of total joint replacement, accumulating to 10–20% revisions after 10 years. The major implant registers report, e.g., for total hip replacement (THR) that of all failures, 50% are attributed to aseptic loosening of the components, 15–18% to dislocations, and 13–18% to wear and osteolysis. It has been reported that about 60% of all patients develop osteolysis after 6–10 years, although it is mostly asymptomatic. At the same time, the register data shows that the survival for THR is reduced by 50% for younger patients and wear debris induced osteolysis after total joint replacement (TJR) is a major limiting factor for the long-term survival of total joint replacement. In the early 1960s Charnley introduced the concept of low friction THR consisting of a stainless steel stem cemented into the femur and a cemented acetabular cup. The metal head of the mono-block component articulated against a cemented PE socket that proved to be more wear resistant than the previously used low friction polymer, and was a major contribution to the great success of THR. Nevertheless, bearings using ultra-high molecular weight polyethylene (PE) as bearing replacement have been shown to be a major factor for the failure of such components due to its limited wear resistance and the inflammatory nature of its wear particles.

Literature indicates that PE wear rates below a certain threshold substantially reduce the incidence of osteolysis and the need for subsequent revision [1, 27]. This threshold is below 0.1 mm PE wear per year and given in values of mg or mm^3/year, assuming an activity of ca one million cycles/year for a patient. Pedometer investigations have shown that several patients independent of their age make more around two million cycles/year, meaning an increase in the distance crossed, between implant surfaces of a factor two and more [26]. Also, there is a trend to joint replacement at earlier stage and an increased aging of the population, both leading to a several fold increase in usage of any implant component. The wear rate of historic PE is around 0.05–0.50 mm/year. Results from retrieval studies

R.M. Streicher
CH-8803 Rüschlikon, Switzerland
e-mail: robert.streicher@stryker.com

K. Knahr (ed.), *Tribology in Total Hip Arthroplasty*,
DOI: 10.1007/978-3-642-19429-0_5, © 2011 EFORT

5

of hard-on-hard articulation such as metal-on-metal or ceramic-on-ceramic show much lower wear rates in the range of 0–0.03 mm/year and lead to their reintroduction to orthopedics to diminish or solve the wear related issues of THRs in the late 1980s and 1990s. Nevertheless, during the past 4 decades and until today metal/ceramic-on-PE articulation is still the gold standard for TJR and yearly approximately two million PE components are implanted worldwide.

Highly cross-linked polyethylene (hxPE) has already been introduced in the late 1970s for THR. Three versions of intentionally highly cross-linked PE have been historically used as cemented cups of THR; one of those types has been chemically cross-linked while the other two were cross-linked using high doses of irradiation, the primary and most reliable means of creating cross-linked PE. None of them used any post-treatment to avoid post-oxidation. Retrospective wear measurements and anecdotally results of >10 year are available: Wroblewski [34] reports in his 10 year data, a wear rate of 0.037 mm/year, a reduction of 75% versus conventionally sterilized PE and Oohnishi [20] as well as Grobbelaar [9] report 20 year results with similar reduction in wear rate. Nevertheless, the historical cross-linking methods used the standard sterilization technology of those days and, therefore, were sub-optimal. Gamma irradiation sterilization in inert gas was introduced in 1986 [28] and since the late 1990s, the second-generation of hxPEs have been used successfully as bearing surfaces using two thermal methods to reduce the free radicals generated during the cross-linking process. Despite a dramatic reduction in wear and positive clinical results with the second generation hxPEs [23], it was necessary to compromise between oxidation resistance and preservation of toughness, which has been addressed by various manufacturers in different ways. Second-generation hxPEs have been used on a limited basis as bearing surfaces for total knee replacement (TKR) and there is limited clinical information on their success but no reports of failures of tibial inserts. Nevertheless, isolated reports of fractures of remelted hxPE hip liners have been reported [11, 24] and degradation due to oxidation remains a long-term concern.

Since 2002 third-generation hxPEs have been developed to minimize the compromise made with second-generation hxPE, the first being introduced in 2005.

5.2
Evolution of PE

The evolution of PE since its introduction in orthopedics is shown in Table 5.1. PE has been introduced Charnley for THR in 1962 looking for a substitution for a low friction polymer he had used earlier. Over the years the quality of PE was improved through enhanced manufacturing methods and rigorous material and product control. A Ca-stearate free quality has become the medical grade industrial standard due to its high purity and homogeneity since 1985 [29].

PE is a linear homopolymer consisting of repeating $-CH_2-$ units with very long and entangled molecular chains. It is a biphasic polymer consisting of crystalline domains in an amorphous matrix. The unique arrangement of ultra-high molecular weight PE with chains to be randomly part of crystallites, responsible for strength and stiffness, the amorphous

Table 5.1 Evolution of UHMWPE for joint implants

Method/technology	Name/example	Reported year of introduction
UHMWPE	RCH-1000, Chirulen	1962
Gamma irradiation sterilization (air)		1968
Carbon fiber reinforcement	Poly-II	1970
First generation highly cross-linked		1972–1978
Higher purity, better consolidation, manufacturing in clean room	Medical grade	1985
Quality without Ca stearate	GUR 402/405	1985
Gamma sterilization in inert gas	Sulene	1986
High-pressure remelted	Hylamer, Hylamer M	1987
Surface heat polishing	PCA	1989
Gamma inert gas sterilization/annealing	Duration	1996
Second generation highly cross-linked, annealed	Crossfire	1998
Second generation highly cross-linked, re-melted	Durasul, Longevity, Marathon	1999–2001
Third generation sequentially highly cross-linked, annealed	X3	2005
Third generation vitamin E doped highly cross-linked, annealed	E1	2007

matrix, responsible for toughness and the entanglements acting as pseudo-cross-links, results in the specific properties of this type of PE. The strong chemical intramolecular bonding and the weak physical intermolecular bonding of the PE molecules make PE sensitive to out of plane cross-shear motion, which is the normal kinematics in the hip joint and to a certain extent in TKR, especially in deeper flexion.

Chemical disinfection was used for PE acetabular components in the early 1960s due to its thermolabile structure and by the late 1960s gamma sterilization to nominal 25 kGy was routinely used. This enabled the components to be sterilized in packaged form. Gamma sterilization in air containing packaging was the predominant method of sterilization until the early 1980s, when questions arose regarding oxidation of components due to the reaction between oxygen and free radicals created during radiation sterilization. Molecular changes are induced during irradiation sterilization causing chain scission, cross-linking, and oxidation, depending on the absorbed dose and the atmosphere in which irradiation takes place. A major improvement in wear resistance and reduced degradation of PE was made in 1986 when the environment for gamma-sterilization of UHMWPE implant components was changed from ambient air to inert gas (Sulene®, Sulzer). This was the first attempt to use the energy of the irradiation sterilization treatment for intentionally cross-linking of the PE and showed a reduction in wear rate of 30% and reduced dramatic fatigue and delamination wear of TKR. This sterilization method was adapted only after 1991 in the US and most major implant manufacturers introduced their variation of this method since then. Additional stabilization of UHMWPE components by a thermal method after sterilization was introduced in 1996 (Duration®, Howmedica). Clinical results, now at 10 years show a significant reduction in wear rate of 35% compared to identical but in air irradiation sterilized PE components [8].

5.3
Highly Cross-Linked PE

HxPE is already in clinical use since 1976 and has shown in anecdotal reports superior wear resistance. Extended cross-linking is accomplished by gamma or electron-beam irradiation of PE bars or sheets with cumulating irradiation doses. As the radiation dose increases the wear resistance is increased (Table 5.2) until an asymptote is reached at about 100 kGy.

Further increase in irradiation does not show any benefit in wear reduction, while the mechanical properties, especially the toughness of hxPE degrade. A subsequent process, either annealing (below the melt temperature) or remelting (above the melt temperature) can reduce or eliminate free radicals from the high energy irradiation process that might else induce oxidation of hxPE in the long-term. This is then followed by a final machining, packaging, and sterilization process, the sterilization generally performed by gas, gas-plasma or conventional gamma sterilization in inert environment. The first new generation hxPE used annealing as radical quenching process and has been introduced in 1998 (Crossfire™, Stryker). Several other brands, all remelted, followed shortly thereafter.

5.3.1
Wear

Highly cross-linked PE demonstrates a dramatic reduction in wear in simulator testing compared to conventionally produced and in inert environment irradiation sterilized PE components. Laboratory data from various research groups and institutes have shown a

Table 5.2 Hip simulator wear rates for UHMWPE following irradiation

Radiation dose (kGy)	Wear rate (mm³/million cycles)
0	140
30	50
50	30
75	10
100	5

reduction in wear of 90 or more percent or less (Wang [32]). The amount of particles is dramatically reduced but their size and morphology for some of the hxPEs is not altered, an important aspect for any histiocytic response. Moreover, independent studies have shown hxPE to be more resistant to wear, even after accelerated aging or when exposed to foreign body debris (Taylor [31]). This generation of hxPE has been widely used in clinical practice for THR but also with restrictions for TKR components. Clinically reductions of 60–80% for THR have been reported in randomized hip replacement studies compared to inert gas gamma-irradiated UHMWPE components and the observation times reach now 10 years. Although one RSA study concluded that the wear rate of specific brands of remelted hxPEs increases after 5 years [14], this has not been confirmed by other clinical studies yet [18] and needs to be followed carefully.

Simulator studies have also shown that due to the direction independence of the hxPEs in contrary to no cross-linked PE the wear of artificial hip joints becomes independent on the head size and thickness of the cup/insert components [13]. This in consequence allows the use of larger heads and thinner components which address anatomical situation and the need to address the increasing dislocation and subluxation incidences, being a major course for early failure of THRs. A larger head will also increase the range of motion and consequently reduce implant/implant impingement. Studies have also shown an enhanced stability sensation with such restored hip joints and consequently enhanced patient satisfaction. Studies, which were conducted to confirm the theoretical advantages of the cross-linking have been conducted by producing various cup inserts made from hxPE and machining various internal diameters into cups of various outer diameters. The resulting PE thickness ranged from 1.8 to 7.9 mm and the inserts were tested up to 5 million cycles at various inclination angels in standard hip joint simulators [30]. Similar tests with similar components have been conducted in other laboratories with deviating test protocols [15]. All results so far confirm that hxPE produces similar low wear rates independent of head size and liner thickness even in impingement mode.

5.3.2
Mechanical Integrity

Despite the well documented advantages of cross-linking and subsequent thermal processes for achieving a dramatic wear reduction, other issues with this category of polymers have been demonstrated or raised. In general, changes in its morphology will compromise the mechanical behavior of hxPEs, such as, e.g., ductility and toughness. While the irradiation source, if ^{60}Co or electron-beam, has not shown different effects, the post irradiation treatment definitely has. Annealing affects the toughness and mechanical resistance to a much lesser extent than remelting, as the PE microstructure is modified to a lesser amount and, therefore the relationship between crystalline and amorphous phases is not changed while some clinical fractures of remelted liners have been reported [11]. Some post-treatments also affect the dimensions of the crystallites which are key for the mechanical response of the hxPE. This reduction of the mechanical resistance has been addressed by several companies by either using lower irradiation doses for hxPE for THR and TKR or only for TKR.

5.3.3
Oxidation

On the other hand, annealing does not eliminate the free radicals completely while remelting does. Although oxidation does not seem to be a limiting factor for THR in the mid-term, and also oxidizing hxPE [33] did not exhibit any implant failure until more than 10 years in vivo [4], there is still concern about long-term results. Because oxidation of UHMWPE takes months or years to reach appreciable levels at ambient or body temperature, thermal aging techniques have been developed to accelerate the oxidation of UHMWPE, with the expectation that the mechanical behavior after accelerated aging will be comparable to naturally aged material [6]. The mechanical behavior of UHMWPE evolves during natural (shelf) aging after gamma irradiation in air, but the kinetics and characteristics of mechanical degradation remain poorly understood, largely due to previous emphasis on indirect measurement techniques. Furthermore, while it is recognized that aging at elevated temperatures will accelerate the oxidation of air-irradiated UHMWPE, the clinical relevance of such a thermally degraded material remains uncertain, particularly if fatigue or joint simulator testing is to be performed after aging. This is even more so in view of the fact that THR does not exhibit increasing wear rates with time, nor is the wear rate well correlated with the shelf aging time. Such an observation suggests that accelerated aging of hip inserts may not reflect either the chemical or clinical pathway actually taken by such inserts, and recent results of oxidizing remelted hxPE support this [25].

5.4
Third-Generation Intentionally Cross-Linked PE

The newest third-generation of hxPE has been introduced in 2005/2007 to address the deficiencies of the previous generation by usage of enhanced technologies to minimize the compromise made with second-generation materials. All new generation hxPEs are now irradiated by gamma-rays instead of electron-beam. To retain the mechanical properties, the annealing procedure is used for all materials on the market, while for the quenching of the free radicals two different methods are use. One uses a sequential irradiation/annealing process repeated three times (X3), others incorporate vitamin E as radical scavenger in various amounts and processes into the hxPE. Both methods are still compromises – the annealing due to the thermodynamics at the temperature chosen to maintain the mechanical properties does not completely eliminate all radicals and some oxidation is still possible, although much below the values achieved with historic hxPEs and inert gas gamma sterilized PE, which have not shown to create a clinical issue even after >20 years of usage, while the incorporation of antioxidants into PE raises some concern.

A battery of various tests have been conducted with sequentially cross-linked hxPE (X3®, Stryker), which is based on its predecessors, the first annealed in inert gas sterilized PE (Duration®) and the first highly cross-linked PE (Crossfire™), which both yielded excellent clinical results after 10 years. These tests included structural analysis, strength determination, free radical concentration, and simulator wear tests of hip joints

Table 5.3 Test battery used for X3

	Property	Test method
Biocompatibility structure		IS0 10993
	Morphology	TEM
	Crystallinity	DSC, SAXS
	Cross-linking density	Swell-ratio
Mechanical integrity (strength, function)	Basic quasi-static tests	
	Functional tests	Fatigue, structural fatigue
		subluxation
		Worst-case scenario
Oxidation (degradation)	Oxygen-bomb resistance	ASTM
	Free radicals	ESR
	Mechanical properties after aging	Various tests
Wear (function)	Vs. control, other designs	Pin-on-disc, hip, and knee simulators
	Wear debris, biological activity	
	Worst-case scenario	Aged, impinged, near impinged, mal-aligned
Clinical behavior	General performance	
	Wear	RSA, Martell

(three institutions) and knee joints (four institutions) before and after oxidative aging (Table 5.3) and benchmarking versus a control group of either standard in inert gas sterilized or non-treated components.

The morphology of the structure of the hxPE X3 is qualitatively and quantitatively similar to the PE used since 20 years, and, therefore the structural integrity and the physical properties of this hxPE are little affected by cross-linking/annealing or aging processes. The mechanical properties are not altered neither by the treatment nor by aging. Even after artificial and real-time aging for 5 years, there is no detectable oxidation and its mechanical properties are not affected [37]. Dynamic fatigue testing showed a survival rate after testing comparable to controls in contrast to remelted hxPEs, which failed up to 100% in this harsh test [5]. Simple tribological tests confirmed the direction independence as well as the delamination resistance of this hxPE and the wear rate is reduced by ca. 90% for hips, and 80% for cruciate retaining as well as posterior stabilized knee implants and the wear particle size and morphology is similar to the standard PE used currently [7]. More challenging testing in impingement mode for THR and maligned mode for TKR confirmed the forgivingness of X3 even in aged conditions [12, 15]. Clinical wear measurement in THR has shown similar results with 14 μm/year in both, a RSA and a radiographic follow up at 2 resp. 3 years follow up [3, 4]. Measurements of retrievals confirmed these low wear rates [19].

The other method applied for the production of third-generation hxPEs is the addition of vitamin E or other radical scavengers [16] adopted from the pharmaceutical, food, chemical and polymer processing industry, introduced to orthopedics since 2007 (E1®, Biomet; ECIMA®, Corin; Vitamys®, Mathys). Several methods have been introduced to blend the antioxidants with PE [17, 22]. Some blend vitamin E with the polymer powder before the consolidation process to produce sheets or bars and cross-link and anneal them, while others diffuse and homogenize the antioxidant using a temperature treatment after cross-linking PE. The sterilization then can be by irradiation or non-irradiation. Both manufacturing methods consume already the antioxidant during either cross-linking or final sterilization. The results reported so far from laboratory [10, 21] and early clinical results for one of the vitamin E doped hxPEs look encouraging [2].

Several concerns with this new technology have been voiced. PE is a paraffin and inclusions or other substances agglomerate at its grain boundaries and consequently affect the mechanical properties, the reason to have eliminated Ca-stearate. Comparative laboratory data has shown that the addition of, e.g., vitamin E to the virgin PE degrades its strength [35, 36]. The other concern is about the antioxidant itself: the optimum method to achieve homogeneity is not clear yet. The ideal amount of antioxidant has not been established yet and it is a non-renewable resource, so may not be sufficient to last for the life time of the patient. Although, e.g., vitamin E is biocompatible by itself this is not evident for any of the reaction products produced during quenching radicals created by the irradiation. Clinical experience will show if these doped hxPEs will come up to expectations in the long-term.

5.5
Summary

Patients' demographics are changing and their expectations are rising. This has a major impact on the bearings used in TJR. Wear and subsequent osteolysis can jeopardize the long-term survival and the best articulation design and materials need to be applied to reduce or avoid its incidence. Ultra-high molecular weight polyethylene has been used as a bearing surface in TJR for over 40 years and the metal/ceramic-on-PE articulation is the gold standard for hip and especially for knee joint implants. On the design side, there is a clear trend to bigger diameters for THR, and for TKR the introduction of high flex implants with an increased rotation movement is challenging current PE performance.

Gamma irradiation sterilization of PE in inert gas introduced 25 years ago, yielded significantly enhanced wear resistance, and reduced aging issue as evident from long-term clinical experience. The introduction of second-generation hxPEs more than 10 years ago using two thermal methods to reduce free radicals generated during the cross-linking process has proven a further reduction in wear rates and yielded positive clinical results. These hxPEs allow for bigger head diameters without increasing the wear and are addressing thus another clinical issue, dislocation, at the same time. Nevertheless, they compromised between oxidation resistance and the preservation of toughness and some drawbacks have been noticed in their clinical application. The third-generation hxPEs minimize the compromise made with second-generation hxPE, and early clinical results confirm their benefits. Highly cross-linked PE is a powerful material to reduce the amount of the wear particles to a sub-risk level and improve the chance for better long-term results of artificial joint prostheses. Careful and diligent follow-up in their clinical application is needed to determine their ultimate success.

References

1. Barrack, R., Folgueras, A.: Pelvic lysis and polyethylene wear at 5–8 years in an uncemented total hip. Clin. Orthop. Relat. Res. **335**, 211–217 (1997)
2. Bragdon, C.R., Greene, M.E., Rubash, H.R., Freiberg, A., Malchau, H.: Early RSA evaluation of wear of vitamin E stabilized highly cross-linked polyethylene and stability of a new acetabular component. Trans. 56th ORS: 2344 (2010)
3. Campbell, D.G., Field, J.R., Callary, S.A.: Second-generation highly cross-linked X3 polyethylene wear: a preliminary radiostereometric analysis study. Clin. Orthop. Relat. Res. **468**(10), 2704–2709 (2010)
4. D'Antonio, J.A., Capello, W.N., Bierbaum, B., Ramakrishnan, R.: Annealed highly cross-linked polyethylenes: clinical performance and second generation materials. Abstr. AAOS: 106 (2010)
5. Dumbleton, J.H., D'Antonio, J.A., Manley, M.A., et al.: The basis for a second-generation highly cross-linked UHMWPE. Clin. Orthop. Relat. Res. **453**, 265–271 (2006)
6. Edidin, A.A., Jewett, C.W., Kwarteng, K., et al.: Degradation of mechanical behavior in UHMWPE after natural and accelerated aging. Biomaterials **21**, 1451–1460 (2000)
7. Essner, A., Yau, S.-S., Schmidig, G., Wang, A., et al.: Reducing hip wear without compromising mechanical strength: a next generation cross-linked and annealed polyethylene. Trans. 5th combined ORS: 80 (2005)
8. Geerdink, C.H., Grimm, B., Vencken, W., et al.: Cross-linked compared with historical polyethylene in THA. Clin. Orthop. Relat. Res. **467**(4), 979–984 (2009)
9. Grobbelaar, C.J.: Longterm results with crosslinked PE. Abstr. 4th EFORT: 158 (1999)
10. Haider, H., Weisenburger, J.N., Kurtz, S., et al.: Highly crosslinked UHMWPE in TKA – does vitamin E-stabilized PE address our concerns? Trans. 55th ORS: 2328 (2009)
11. Harris, W.: Three revolutions in acetabular revision surgery. Abstr. 16th ISTA: 24–28 (2003)
12. Hermida, J.C., Fischler, A., Colwell, C.W., D'Lima, D.D.: The effect of oxidative aging on the wear performance of highly crosslinked polyethylene knee inserts under conditions of severe malalignment. J. Orthop. Res. **26**(12), 1585–1590 (2008)
13. Herrera, L., Lee, R., Longaray, J., et al.: Hip simulator evaluation of the effect of femoral head size on sequentially cross-linked acetabular liners. Wear **263**, 1034–1037 (2007)
14. Johanson, P.-E., Digas, G., Thanner, J., et al.: Clinical performance of highly cross-linked polyethylene with seven years follow-up. Trans. 56th ORS: 357 (2010)
15. Kelly, N.H., Rajadhyaksha, A.D., Wright, T.M., et al.: High stress conditions do not increase wear of thin highly crosslinked UHMWPE. Clin. Orthop. Relat. Res. **468**, 418–423 (2009)
16. King. R., Narayan, V.S., Ernsberger, C., Hanes, M.: Characterization of gamma-irradiated UHMWPE stabilized with a hindered-phenol antioxidant. Trans. 55th ORS: 19 (2009)
17. Leer, R., Zurbruegg, D., Delfosse, D.: Use of vitamin E to protect cross-linked UHMWPE from oxidation. Biomaterials **31**, 3643–3648 (2010)
18. Martell, J., Clohisy, J., White, R., et al.: Multi-center study of the mid-term follow-up results of highly cross-linked polyethylene THR components. Trans. 56th ORS: 359 (2010)
19. MacDonald, D., Sakona, A., Austin, M., Parvizi, J., Kurtz, S.M.: Comparison of first – and second-generation annealed polyethylenes for total hip arthroplasty. Trans. 56th ORS: 2351 (2010)

20. Oonishi, H.: Long term clinical results of THR. Clinical results of THR of an alumina head with a cross-linked UHMWPE cup. Orthop. Surg. Traumatol **38**, 1255–1264 (1995)
21. Oral, E., Christensen, S.D., Mahlhi, A.S., Wannomae, K.K., Muratoglu, O.K.: Wear resistance and mechanical properties of highly cross-linked UHMWPE doped with vitamin E. J. Arthroplasty **21**(4), 580–521 (2006)
22. Oral, E., Wannomae, K.K., Rowell, S.L., Muratoglu, O.K.: Diffusion of vitamin E in UHMWPE. Biomaterials **28**, 5225–5237 (2007)
23. Roehrl, S.M., Li, M.G., Nilsson, K.G., Nivbrant, B.: Very low wear of non-remelted highly cross-linked polyethylene cups: An RSA study lasting up to 6 years. Acta Orthop. **78**(6), 739–745 (2007)
24. Rowell, S.L., Duffy, G., Wannomae, K.K., et al.: Failure of a Marathon™ UHMWPE acetabular liner: a case study. Trans. 53rd ORS: 1646 (2007)
25. Rowell, S.L., Yabannavar, P., Muratoglu, O.K.: Oxidative stability of simulator-tested acetabular liners after 7-years shelf aging in air. Trans. 56thORS: 358 (2010)
26. Schmalzried, T.P., Szuszczewicz, E.S., Northfield, M.R., et al.: Quantitative Assessment of Walking Activity after Total Hip or Knee Replacement. J. Bone Joint Surg. **80A**, 54–59 (1998)
27. Shih, C.-H., Lee, P.-C., et al.: Measurement of polyethylene wear in cementless total hip arthroplasty. J. Bone Joint Surg. **79-B**, 361–365 (1997)
28. Streicher, R.M.: Investigation on sterilization and modification of high molecular weight polyethylenes by ionizing irradiation. Beta Gamma **89**(1), 34–43 (1989)
29. Streicher, R.M.: UHMWPE as the substance for articulating components of joint prostheses. Biomed. Tech. **38**(12), 303–313 (1993)
30. Streicher, R.M., Wang, A.: Mechanical and tribological properties of cross-linked UHMWPE. Riv. patologia 'apparato locomotore **Vol II**(1, 2), 39–46 (2003)
31. Taylor, S., Serekian, P., Bruchalski, P., Manley, M.T.: The performance of irradiation-cross-linked UHMWPE cups under abrasive conditions throughout hip simulation testing. Trans. 45thORS: 252 (1999)
32. Wang, A., Manley, M.T., Serekian, P.: Wear and structural fatigue simulation of crosslinked ultra-high molecular weight polyethylene for hip and knee bearing applications. ASTM STP **1445**, 151–168 (2003)
33. Willie, B.M., Bloebaum, R.D., Ashrafi, S., et al.: Oxidative degradation in highly cross-linked and conventional polyethylene after 2 years of real time shelf aging. Biomaterials **27**, 2275–2284 (2006)
34. Wroblewski, B.M., Siney, P.D., Fleming, P.A.: Low-friction arthroplasty of the hip using alumina ceramic and cross-linked polyethylene. J. Bone Joint Surg. **81B**(1), 54–55 (1990)
35. Yau, S.S., Wang, A., Lovell, T.: Evidence of compromised performance properties in vitamin E-doped irradiated UHMWPE. Abstr. 8th World Biomaterials Congress: 1691 (2008)
36. Yau, S.S., Le, K.-P., Wang, A.: Doping vitamin E by diffusion deteriorates properties of cross-linked UHMWPE: verified by experiments. Trans. 55th ORS: 458 (2009)
37. Yau, S.-S., Le, K.-P., Blitz, J.W., Dumbleton, J.H.: Real-time shelf aging of sequential cross-linked and annealed UHMWPE. Trans. 55th ORS: 450 (2009)

Part II

Ceramic Articulations

Are Noisy Ceramic-on-Ceramic Hips Linked to Periprosthetic Bone?

6

Bernd Grimm, Alphons Tonino, and Ide Christiaan Heyligers

Abbreviations

CoC   Ceramic-on-ceramic
CWT   Cortical wall thickness
DEXA Dual energy x-ray absorptiometry
LT   Lesser trochanter
PPB   Periprosthetic bone

6.1
Introduction

The number of reports published about noisy, mainly squeaking but also scratching or clicking ceramic-on-ceramic (CoC) total hip arthroplasties (THA) has been increasing rapidly during the past few years. At the same time, the reported incidence rates of noisy CoC hips vary widely between 0% (0/177) [1], 0.3% (1/301) [2], 4.8% (8/168) [3], 4.9% (5/97) [4], 6.4% (95/1486) [24] and for instance 10.7% (14/131) [5] and up to 20.3% (9/43) [6] or even 35.6% (16/45) [7] for a certain implant design. These differences can be explained by the method of investigation (e.g., pro-actively reporting patients only versus thorough individual patient interviews), the type(s) of noise recorded (squeaking, clicking/popping/snapping, scratching/grinding), by setting the threshold of clinical relevance (e.g., frequency of occurrence, audibility to others, type of associated movement and relevance in activities of daily living) or the subjectivity inherent with any questionnaire based patient assessment.

While it has been shown in one study [8] that the patient's experience of some CoC noise is not uncommon (41/362 = 11.3%) only a small proportion (2/362 = 0.6%) reported audible squeaking and frequent occurrence. While the association of squeaking with pain,

B. Grimm (✉), A. Tonino, and I.C. Heyligers
Department of Orthopaedic Surgery & Traumatology, Atrium Medical Center,
AHORSE Research Foundation, Henri Dunant-Str. 5, 6419PC Heerlen, The Netherlands
e-mail: b.grimm@atriummc.nl

K. Knahr (ed.), *Tribology in Total Hip Arthroplasty*,
DOI: 10.1007/978-3-642-19429-0_6, © 2011 EFORT

6

lower hip scores or less quality of life was either absent [4, 8, 24, 25] or low at 0.9% (3/320) [9] other studies have reported revisions due to squeaking [6, 10]. Even when the clinically critical incidence of squeaking hips is low it remains a phenomenon linked to this particular bearing option which needs further understanding to be reduced or eliminated by new materials, designs, surgical techniques or patient selection.

Several studies were able to identify surgery, patient or implant related risk factors associated with the development of squeaking hips. Most consistently the orientation of the acetabular cup outside the Lewinnek's "safe zone" [11], in particular a too steep cup inclination has been associated with a higher chance of developing noise [12]. Other factors commonly reported to enhance the chance of a squeaking CoC bearing seem to be related to joint laxity [10, 13], such as the use of short neck stems [6], leg length correction or the surgical approach [10]. Other factors which were described in some studies but not found in others or even reported as contraindicators are weight, height, BMI, gender, or activity levels [3, 10, 14]. It has become obvious that the squeaking phenomenon is multifactorial with an etiology not yet completely understood.

The etiology seems to always involve stripe wear [15] resulting from micro-separation, impingement, edge loading, third particle ingress, metal transfer or fluid film disruption, producing a stick-slip effect in the bearing which excites vibrations [16, 17]. The fact that stripe wear is conditional to develop CoC squeaking is also used to explain the relatively late onset of squeaking after a mean of 14 months [12] to 26.4 months [6] as a certain in vivo use is required to develop these particular wear scars.

As stripe wear is also found in silent CoC bearings, a theory has been developed that the vibrations become audible only via amplification through the vibrating stem. This was supported by showing that the excitation frequency and the resonance frequency of the plain stem are similar [18, 19]. This theory has also been used to explain apparently different incidence rates of squeaking between different stem designs and metallurgy, parameters which influence the resonance behavior [18]. However, stem resonance in vivo would also be influenced by the periprosthetic bone damping and transmitting stem vibrations. Thus, if the resonating stem theory were true, noisy CoC hips should show periprosthetic bone different to silent hips.

This study compares stem fit&fill and periprosthetic bone between noisy and silent CoC hips.

6.2
Patients and Methods

In a consecutive series of 186 primary total hip arthroplasties with a CoC bearing using identical stems, cups (Stryker ABG-II) and femoral heads (Alumina with V40 taper, 28 mm head diameter), a dedicated survey was performed during the regular annual follow-up visit to identify patients experiencing noisy hips. The questionnaire investigated the incidence of any noise, the type of noise (squeaking, clicking, scratching, combinations), the frequency of occurrence (often, sometimes, rare), the level of noise (audible by others, self-audible, reproducible) the movement types linked to noise (walking, stair climbing,

Table 6.1 Patient characteristics of the noisy group and the matched control group

Patient characteristic	Noisy group	Silent group
Number	38	38
Gender female/male	23/15	23/15
Mean age (years)	53.3±7.9 (36–65)	53.2±5.8 (36–66)
Mean follow-up time	4.6 years	4.6 years
Stem size	3.47	3.45

chair rising, bending, lifting) and the time of onset. The questionnaire identified 38 noisy hips resulting in an incidence rate of 20.4%. The sub-group of squeaking hips counted $n=23$ patients (12.4%). From the remaining 148 silent hips, a control group was selected matching patients for gender, age, follow-up time and stem size (Table 6.1).

The endosteal canal width, the stem "fit & fill" and cortical wall thickness (CWT, medial and lateral) were measured digitally on the last available post-operative antero-posterior radiograph (Dicom, $2,494 \times 2,048$ pixels) according to an established method [20, 26] using the Roman V1.7 freeware program [21] and image calibration on the 28 mm femoral head. Based on a central line through the endosteal canal, three measurement levels perpendicular to the central line are drawn (Fig. 6.1). The distal level is 10 mm proximal to the tip of the stem. The mid-stem level is half the distance between the tip and shoulder of the stem and usually below the lesser trochanter. When the mid-stem level cuts through the lesser trochanter (LT), then the mid-stem level is shifted slightly distal to the distal end of the LT. While the distal level is always and mid-stem level is most frequently stem referenced, the proximal level is bone referenced and set at the proximal end of the lesser trochanter. The "fit & fill" is defined as the ratio in percent between the distance covered by the metal stem and the endosteal canal width at the particular level (Fig. 6.2).

Measurements were repeated by a single blinded observer in a control group of silent hips matched for gender, age, stem size and follow-up time (mean=4.6years). Fit & fill and CWT were compared between the noisy and silent group at proximal, mid-stem and distal level and on the medial and lateral side. Groups were compared using the student t-test after verification of a normal distribution by applying the Kolmogorov–Smirnov test. A conventional significance level was set at $p=0.05$.

6.3
Results

The endosteal canal width was equal in noisy (N) and silent hips (S) at all three measured levels (e.g. proximal: $N=39.7\pm5.5$ mm, $S=41.3\pm5.7$ mm, $p>0.05$, see Fig. 6.3). On the lateral side also cortical wall thickness (CWT) was the same at all three levels (e.g. proximal: $N=2.0\pm0.8$ mm, $S=1.8\pm0.9$ mm, $p>0.05$, see Fig. 6.4). However, on the medial side, noisy hips had a significantly higher CWT at proximal ($N=4.9\pm2.8$ mm,

6

Fig. 6.1 The stem and bone related measurement levels (proximal, mid-stem, distal) according to method of Kim and Kim (Kim et al. 1993)

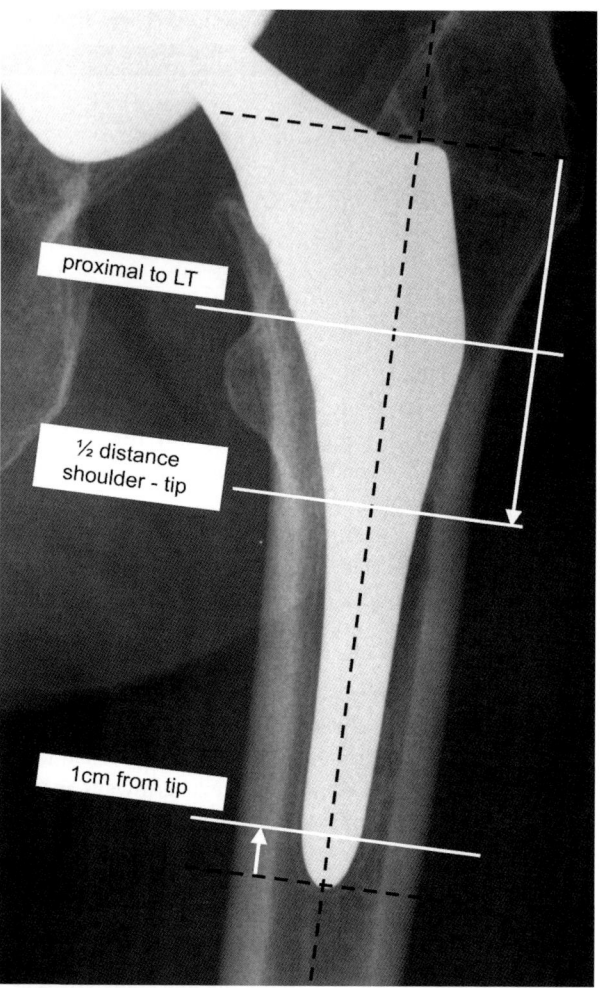

Fig. 6.2 Example of the measurements taken at the distal stem level: *e* endosteal canal width, *s* stem thickness, c_M cortical wall thickness medial, c_L cortical wall thickness lateral

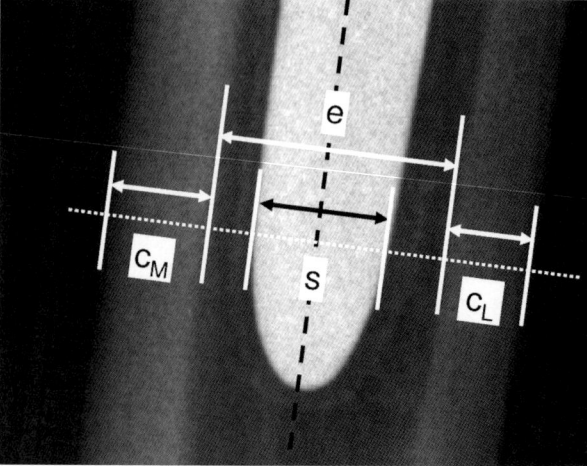

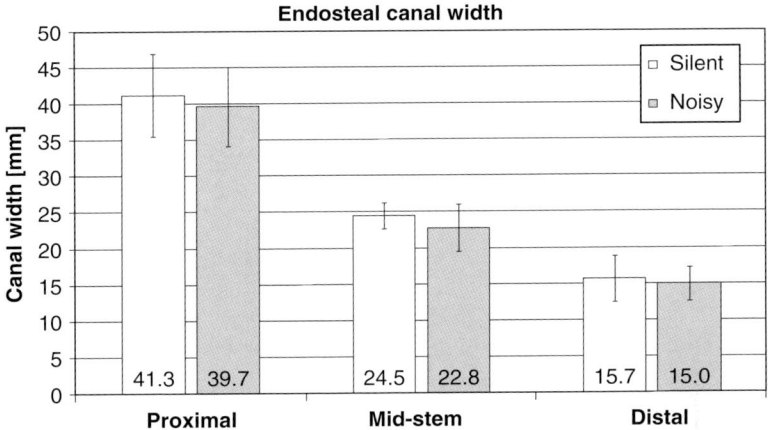

Fig. 6.3 The endosteal canal width (mm) for the three stem/bone levels compared between the silent and the noisy group

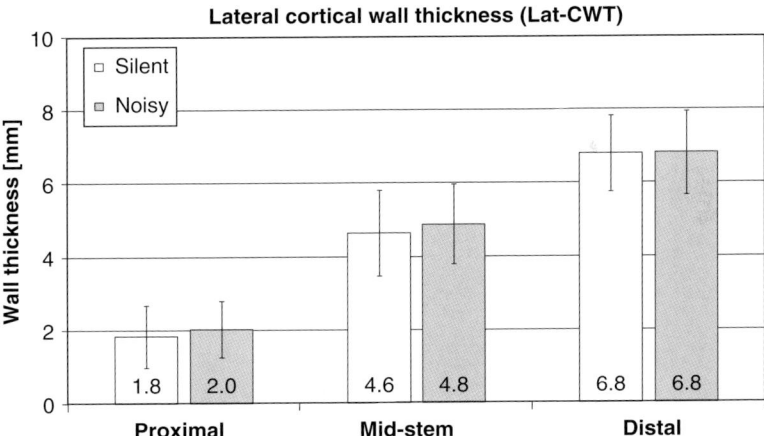

Fig. 6.4 The lateral side cortical wall thickness (mm) for the three stem/bone levels compared between the silent and the noisy group

$S = 3.0 \pm 2.1$ mm, $p < 0.01$) and mid-stem level ($N = 6.2 \pm 2.1$ mm, $N = 4.6 \pm 1.7$ mm, $p < 0.001$, Fig. 6.5). Also "fit & fill" was slightly higher at proximal level ($N = 66\%$, $S = 62\%$, $p < 0.05$) and mid-stem level ($N = 63\%$, $S = 59\%$, $p < 0.05$, Fig. 6.6). Differences and significance levels increased when in the noise group only squeakers were considered (Fig. 6.5).

The proportion of stems where the mid-stem line had to moved slightly distally to not cut through the lesser trochanter was twice as high in the noisy group (18/38) an in the silent group (9/38, $p = 0.02$, Fisher exact test). This indicates that there was a tendency of noisy stems to sit more proximal with reference to the lesser trochanter.

6

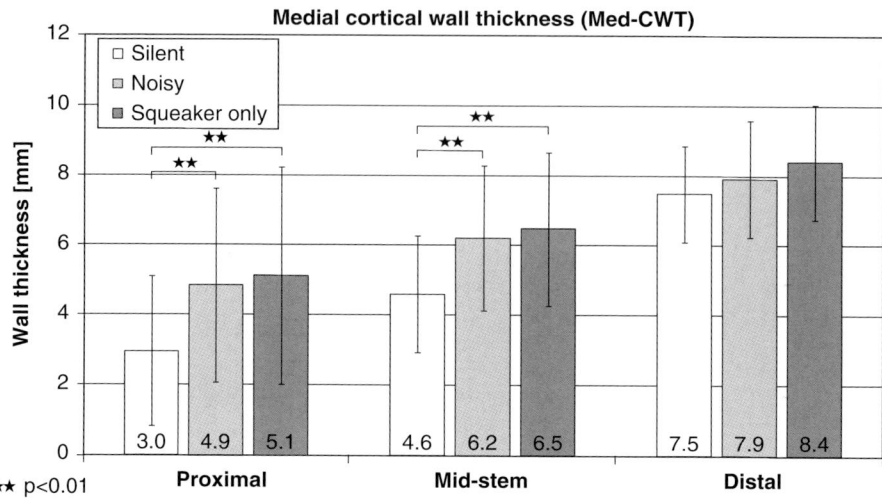

Fig. 6.5 The medial side cortical wall thickness (mm) for the three stem/bone levels compared between the silent group, the noisy group and the squeakers group

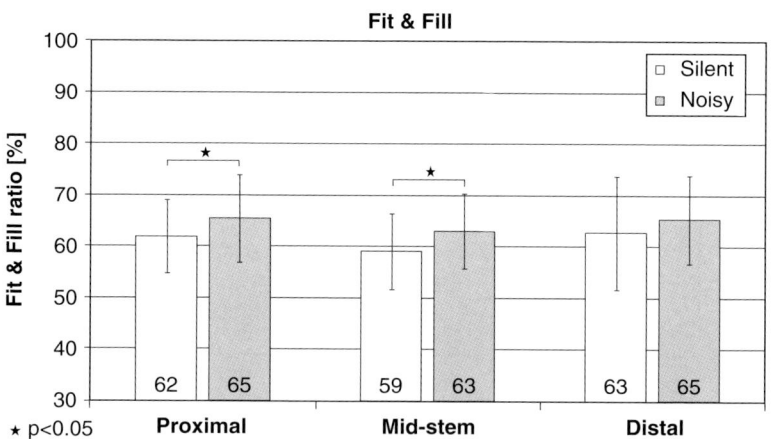

Fig. 6.6 The "fit & fill" for the three stem/bone levels compared between the silent and the noisy group

6.4
Discussion

Noisy, in particular squeaking ceramic-on-ceramic total hips have been reported at incidence rates up to 35.6% [7] for a certain design (Restrepo et al. 2010). Although the incidence rates are lower with other designs and even smaller proportions of patients consider squeaking as problematic or worthwhile proactively reporting it to their doctors, in some cases the psychological effects and social embarrassment have lead to revision [6, 7, 10]. In addition, even low noise volumes and less frequent noise events may be an early warning sign of bearing damage that may lead to late implant failure [22]. Thus, the phenomenon requires further investigation to clarify its etiology and possibly derive advice for the surgeon using current designs (e.g. patient indication, implant selection, surgical procedure) or the designer of future CoC devices to reduce or eliminate the problem.

Audible squeaking starts as a vibration excited at the CoC bearing interface which as a result of micro-separation, sub-luxation, impingement, edge loading or third particle ingress has developed surface changes in the form of stripe wear or metal transfer and is undergoing fluid film disruptions. These inhomogeneous tribological conditions can excite bearing vibration during a motion cycle. As the particular surface changes have been observed also in retrieved silent hips, it has been suggested that the vibration becomes audible only when amplified via a resonating stem and in fact the squeaking frequencies are in the range of the resonance frequencies of typical stems [19]. If stem resonance plays a major role in the development of audible noise, then the periprosthetic bone shall be different between noisy and silent hips as the periprosthetic bone defines the boundary conditions of mechanical support and damping modulating the emitted sound. Thus it was the purpose of this study to compare the periprosthetic bone between noisy and silent hips using an established method of assessing fit&fill and cortical wall thickness on standard radiographs.

Equal endosteal canal width and equal stem sizes between the silent and noisy group and between the silent and squeaking sub-group were found and indicate that matching was successful. It also indicates that stem sizing in relation to the endosteal canal as a pre-operative planning exercise or a per-operative surgical choice was probably not a contributing factor in the development of patient reported noise.

However, noisy hips had significantly thicker medial cortical walls than silent hips at proximal level and at mid-stem level. At the same time the "fit & fill" was slightly higher at both these levels. There was also a tendency for the noisy stems to sit slightly more proximal than the silent stems. This gives evidence that the periprosthetic bone may play a role in the development of noisy CoC hips providing particular conditions of stem anchorage, stem support and damping, all of which may influence the transmission properties of a vibrating stem. From this study it appears that in the proximal to mid-stem region strong medial walls, a tight fit (two normally not negative but even desired properties in the fixation of uncemented stems) and a proximal stem position enhance vibration transmission in such a way that audible noise and in particular squeaking may develop.

There was no difference between silent and noisy hips at distal stem level neither with regards to the canal width, the medial or lateral cortical wall thickness nor the "fit & fill". This is further evidence for the successful patient matching procedure, the equal pre- or per-operative sizing procedures but also for the fact that with regards to vibration transmission and noise development the proximal bone and not the distal bone is the periprosthetic region relevant for stem resonance. Like the distal periprosthetic bone, also the lateral periprosthetic bone showed no difference between silent and noisy hips and thus seems less influential in vibration transmission.

6

Interpreting these findings with regards to clinical consequences one may conclude that certain proximal femora may anatomically be more prone to noise development than others. However, it would be too early to derive criteria for patient selection based on the first data presented in this study.

Considering the equal canal width, the equal lateral wall thickness and the choice of equal stem sizes while mainly the proximal lateral walls were different, it seems like that not the anatomy but surgical canal preparation may have created the particular conditions found. Canal reaming or broaching can lead to asymmetric medial to lateral removal of bone. In addition, the more proximal stem position found in noisy hips may indicate differences in the insertion technique and depth. Similar to thin necks which seem to be more prone to resonate, a more proximally placed stem may also enhance noise amplification. This proximal stem position may be correlated with attempts to perform leg length corrections, a factor previously identified as a risk factor for noise development. The fact that equal canal widths and stem sizes produced differences in antero-posterior fit & fill seems as a contradiction at first but can be explained by different stem rotations. Different stem version may alter stem anchorage and thus vibration transmission but it can also contribute to sub-luxation, the original cause of stripe wear development.

The fact that differences and significance levels between silent and noisy hips increased when only the squeaking sub-group was considered in the comparison further strengthens the notion that audible vibration is influences by the periprosthetic bone. In a further sub-group analysis where short follow-up x-rays were available it appeared that the different trends between silent and noisy hips are present already at short follow-up when major bone remodeling has not taken place yet. This indicates that the conditions enhancing the development of audible noise are either present as an anatomical feature or the result of surgical canal preparation and stem insertion.

This study presents evidence that the periprosthetic bone can play a role in the development of audible CoC bearing noise. This supports the theory postulating that the vibration excited in the bearing requires amplification through an oscillating stem. The results of this study would also explain why CoC hips with clear signs of stripe wear could be silent while others developed squeaking.

A proximal and mid-stem femur with strong medial cortical walls and tight "fit & fill" enhanced resonance as well as a more proximally positioned stem. The results of this study suggest that these particular conditions are not pre-defined by the patient's anatomy but the result of surgical canal preparation and the insertion depth and version of the stem.

However, considering the following study limitations it appears too early to conclude surgical advice as to reduce squeaking based on the presented data only. The particular periprosthetic bone conditions reported in this study may be specific to the anatomic implant design used and thus may not be generalized. It is likely that a straight stem, a multi-tapered design or stems with surface properties different to the proximally coated sleeve of an anatomic stem in this study may exhibit different periprosthetic bone conditions which enhance noise via vibration transmission. Nevertheless it can be expected that periprosthetic bone will also play some role in development of noise in other stem designs.

Besides the issue of generalization, the study is limited by the rather poor inter-rater reliability of measuring periprosthetic bony dimensions in general and also in particular with the method of Kim and Kim [23]. However, while the absolute numerical figures may change upon re-measurement by the same or a different observer, the trends towards the significant relative differences between the groups identified in this study shall remain and this is where the conclusions of this study are based upon.

Another study limitation is that all noisy hips were combined in one group including the squeakers and those with other forms of noise. However, the significant differences between the silent and the overall noisy group were also found when only the squeakers were compared to the silent controls. The fact that also hips with noise other than squeaking exhibit the periprosthetic bone typical for the resonance of squeaking noise cannot yet be fully explained.

Other study limitations were the measurement of only a single, namely the last follow-up radiograph but not a pre-operative radiograph nor a direct or short-term post-operative x-ray. Thus it cannot be verified entirely as to whether the periprosthetic bone conditions found were present already at pre-operative or the result of a bone remodeling process. In addition, the selection of a well matched control group is a scientifically sound and valid approach but measuring all silent hips may have further enhanced the statistical power.

Considering the identified study limitations, in a future investigation it is planned to repeat the measures in a second observation by the same observer and a third observation by another observer. When available, pre-operative and a direct post-operative radiographs shall be included and the entire silent group shall be measured. In addition, the noisy group and the matched silent group shall be measured using DEXA analysis to gain insight into the periprosthetic bone not only with regards to morphometric features like in this study but also with regards to the bone quality such as the bone mineral density.

In conclusion, this study has provided evidence that in the etiology of audible noise in CoC hip bearings the periprosthetic bone can play a role. This confirms theories about the need for a resonating stem to make a vibrating bearing audible. While the findings cannot yet be used to provide advice to surgeons or designers on how to avoid noisy CoC hips, it confirms the highly multifactorial nature of the phenomenon. Thus care shall be taken with attributing squeaking to mainly implant design only.

References

1. Hamilton, W.G., McAuley, J.P., Dennis, D.A., Murphy, J.A., Blumenfeld, T.J., Politi, J.: THA with Delta ceramic on ceramic: results of a multicenter investigational device exemption trial. Clin. Orthop. Relat. Res. **468**(2), 358–366 (2010a). doi:10.1007/s11999-009-1091-4
2. Lusty, P.J., Tai, C.C., Sew-Hoy, R.P., Walter, W.L., Walter, W.K., Zicat, B.A.: Third-generation alumina-on-alumina ceramic bearings in cementless total hip arthroplasty. J. Bone Joint Surg. Am. **89**(12), 2676–2683 (2007). doi:10.2106/JBJS.F.01466
3. Choi, I.Y., Kim, Y.S., Hwang, K.T., Kim, Y.H.: Incidence and factors associated with squeaking in alumina-on-alumina THA. Clin. Orthop. Relat. Res. **468**(12), 3234–3239 (2010b). doi:10.1007/s11999-010-1394-5
4. Greene, J.W., Malkani, A.L., Kolisek, F.R., Jessup, N.M., Baker, D.L.: Ceramic-on-ceramic total hip arthroplasty. J. Arthroplasty **24**(6 Suppl), 15–18 (2009). doi:10.1016/j.arth.2009.04.029
5. Jarrett, C.A., Ranawat, A.S., Bruzzone, M., Blum, Y.C., Rodriguez, J.A., Ranawat, C.S.: The squeaking hip: a phenomenon of ceramic-on-ceramic total hip arthroplasty. J. Bone Joint Surg. Am. **91**(6), 1344–1349 (2009). doi:10.2106/JBJS.F.00970

6. Keurentjes, J.C., Kuipers, R.M., Wever, D.J., Schreurs, B.W.: High incidence of squeaking in THAs with alumina ceramic-on-ceramic bearings. Clin. Orthop. Relat. Res. **466**(6), 1438–1443 (2008). doi:10.1007/s11999-008-0177-8

7. Swanson, T.V., Peterson, D.J., Seethala, R., Bliss, R.L., Spellmon, C.A.: Influence of prosthetic design on squeaking after ceramic-on-ceramic total hip arthroplasty. J. Arthroplasty **25**(6 Suppl), 36–42 (2010). doi:10.1016/j.arth.2010.04.032

8. Schroder, D., Bornstein, L., Bostrom, M.P., Nestor, B.J., Padgett, D.E., Westrich, G.H.: Ceramic-on-ceramic total hip arthroplasty: incidence of instability and noise. Clin. Orthop. Relat. Res. **469**(2), 437–442 (2011). doi:10.1007/s11999-010-1574-3

9. Mai, K., Verioti, C., Ezzet, K.A., Copp, S.N., Walker, R.H., Colwell Jr., C.W.: Incidence of 'squeaking' after ceramic-on-ceramic total hip arthroplasty. Clin. Orthop. Relat. Res. **468**(2), 413–417 (2010). doi:10.1007/s11999-009-1083-4

10. Grimm, B., Tonino, A.J., Vencken, W., Heyligers, I.C.: Noisy ceramic bearings: high incidence rate and the influence of factors affecting joint laxity. Podium No: 199. Am. Acad. Orthop. Surg. http://www3.aaos.org/education/anmeet/anmt2009/podium/podium.cfm?Pevent=199 (2009). Accessed 3 Nov 2010

11. Lewinnek, G.E., Lewis, J.L., Tarr, R., Compere, C.L., Zimmerman, J.R.: Dislocations after total hip-replacement arthroplasties. J. Bone Joint Surg. Am. **60**(2), 217–220 (1978)

12. Walter, W.L., O'toole, G.C., Walter, W.K., Ellis, A., Zicat, B.A.: Squeaking in ceramic-on-ceramic hips: the importance of acetabular component orientation. J. Arthroplasty **22**(4), 496–503 (2007). doi:10.1016/j.arth.2006.06.018

13. Glaser, D., Komistek, R.D., Cates, H.E., Mahfouz, M.R.: Clicking and squeaking: in vivo correlation of sound and separation for different bearing surfaces. J. Bone Joint Surg. Am. **90**(Suppl 4), 112–120 (2008). doi:10.2106/JBJS.H.00627

14. Walter, W.L., Waters, T.S., Gillies, M., Donohoo, S., Kurtz, S.M., Ranawat, A.S., Hozack, W.J., Tuke, M.A.: Squeaking hips. J. Bone Joint Surg. Am. **90**(Suppl 4), 102–111 (2008). doi:10.2106/JBJS.H.00867

15. Nevelos, J., Ingham, E., Doyle, C., Streicher, R., Nevelos, A., Walter, W., Fisher, J.: Microseparation of the centers of alumina-alumina artificial hip joints during simulator testing produces clinically relevant wear rates and patterns. J. Arthroplasty **15**(6), 793–795 (2000). doi:10.1054/arth.2000.8100

16. Chevillotte, C., Trousdale, R.T., Chen, Q., Guyen, O., An, K.N.: The 2009 Frank Stinchfield Award: "Hip squeaking": a biomechanical study of ceramic-on-ceramic bearing surfaces. Clin. Orthop. Relat. Res. **468**(2), 345–350 (2010). doi:10.1007/s11999-009-0911-x

17. Weiss, C., Gdaniec, P., Hoffmann, N.P., Hothan, A., Huber, G., Morlock, M.M.: Squeak in hip endoprosthesis systems: an experimental study and a numerical technique to analyze design variants. Med. Eng. Phys. **32**(6), 604–609 (2010). doi:10.1016/j.medengphy.2010.02.006

18. Hothan, A., Huber, G., Weiss, C., Hoffmann, N., Morlock, M.M.: Femoral stems are decisive components for the characteristics of squeaking. Annual Meeting of the Orthopedic Research Society. Oral presentation 0236. http://ors.org/web/Transactions/56/0236.pdf (2010). Accessed 3 Nov 2010

19. Taylor, S., Manley, M.T., Sutton, K.: The role of stripe wear in causing acoustic emissions from alumina ceramic-on-ceramic bearings. J. Arthroplasty **22**(7 Suppl 3), 47–51 (2007). doi:10.1016/j.arth.2007.05.038

20. van der Wal, B.C., de Kramer, B.J., Grimm, B., Vencken, W., Heyligers, I.C., Tonino, A.J.: Femoral fit in ABG-II hip stems, influence on clinical outcome and bone remodeling: a radiographic study. Arch. Orthop. Trauma. Surg. **128**(10), 1065–1072 (2008). doi:10.1007/s00402-007-0537-y

21. Geerdink, C.H., Grimm, B., Vencken, W., Heyligers, I.C., Tonino, A.J.: The determination of linear and angular penetration of the femoral head into the acetabular component as an assessment of wear in total hip replacement: a comparison of four computer-assisted methods. J. Bone Joint Surg. Br. **90**(7), 839–846 (2008). doi:10.1302/0301-620X.90B7.20305

22. Toni, A., Traina, F., Stea, S., Sudanese, A., Visentin, M., Bordini, B., Squarzoni, S.: Early diagnosis of ceramic liner fracture. Guidelines based on a twelve-year clinical experience. J. Bone Joint Surg. Am. **88**(Suppl 4), 55–63 (2006). doi:10.2106/JBJS.F.00587
23. Kim, Y.H., Kim, J.S., Oh, S.H., Kim, J.M.: Comparison of porous-coated titanium femoral stems with and without hydroxyapatite coating. J. Bone Joint Surg. Am. **85-A**(9), 1682–1688 (2003)
24. Restrepo, C., Matar, W.Y., Parvizi, J., Rothman, R.H., Hozack, W.J.: Natural history of squeaking after total hip arthroplasty. Clin. Orthop. Relat. Res. **468**(9), 2340–2345 (2010a). doi:10.1007/s11999-009-1223-x
25. Restrepo, C., Post, Z.D., Kai, B., Hozack, W.J.: The effect of stem design on the prevalence of squeaking following ceramic-on-ceramic bearing total hip arthroplasty. J. Bone Joint Surg. Am. **92**(3), 550–557 (2010b). doi:10.2106/JBJS.H.01326
26. Kim, Y.H., Kim, V.E.: Uncemented porous-coated anatomic total hip replacement. Results at six years in a consecutive series. J Bone Joint Surg Br. **75**(1), 6–13 (1993)

Noise Emissions in Total Hip Replacements, with an Emphasis on Ceramic-on-Ceramic and Ceramic-on-Metal Bearings and Different Articular Sizes

7

Dick Ronald van der Jagt, Lipalo Mokete, Bradley Rael Gelbart, Kingsley Nwokeyi, and Anton Schepers

The introduction of hard on hard high performance bearing surfaces in total hip arthroplasties has promised improved longevity in respect of the bearing surface, leading to lower anticipated revision rates. A further advantage of these bearing surfaces is that one can considerably increase head size diameter, providing better stability without the risk of greater wear. The strength and durability of ceramics has been improved by the development of an alumina matrix composite (Biolox delta; Ceramtec AG) which consists of 82% alumina, 17% zirconia and 0.3% chromium oxide. This material promises improved wear performance as a like-on-like bearing surface as well as better mechanical properties resulting in a lower risk of breakage [1]. Other improvements in manufacturing techniques include smaller grain size tolerances as well as the lack of inclusions and grain boundaries. Manufacturing tolerances have also improved resulting in better matching at the morse tapers used in these modular prostheses. All these improvements in the materials and manufacturing techniques have been matched with parallel improvements in design, leading to better clinical results.

All bearing surfaces do present with varying complications, but squeaking in ceramic-on-ceramic (CoC) bearings has been identified as a potentially major complication. Alumina CoC bearings were first implanted in the 1970s [2], and noises from these hip bearings were reported early on. These squeaks are audible noises that occur during movement of the hip. They are distinct from other noises that include clunking, clicking, popping and grinding. It should be noted that each and every bearing surface whether used in a medical or in a commercial industrial context will emit noises. These noises may well be outside our normal auditory range, but they do occur. These other sounds have been recorded with acoustic transducers and they include snapping, knocking, cracking, grinding and snap like noises [3]. Others have been described as "thud-like clicking" and "clear and rich clicking" noises. These sounds are associated with movement of the joints in the clinical situation but their mechanisms of generation are poorly understood. While these

D.R. van der Jagt (✉), L. Mokete, B.R. Gelbart, K. Nwokeyi, and A. Schepers
Division of Orthopaedic Surgery, University of the Witwatersrand,
Johannesburg, Morningside, 2057, South Africa
e-mail: dvdjagt@mweb.co.za

K. Knahr (ed.), *Tribology in Total Hip Arthroplasty*,
DOI: 10.1007/978-3-642-19429-0_7, © 2011 EFORT

noises may be disconcerting to the patient, their clinical significance has not been confirmed. They may well have extra-articular origins with little influence on the prosthesis or its longevity.

Some patients are tolerant of these noises, but they may be loud and disconcerting. Clinicians are not sure of the clinical significance of squeaking hips and whether there is a good reason to revise them [4–7]. Definite guidelines are not agreed upon as to whether squeaking bearings should be revised, and the decision to do so is often prompted by other unscientific reasons. Some patients are not prepared to put up with the noise any longer because they find them socially unacceptable. Other patients are influenced by aggressive marketing campaigns from rival orthopedic implant companies. Of note is that when revisions are done for noise, retrieved ceramic components nearly always demonstrate some stripe wear, resulting from microseparation.

The incidence of squeaks and other noises in ceramic articulations varies tremendously between different reported series. When routine follow-ups are used as the basis to determine the incidence of articular squeaks, these are often artificially low because of patients' reluctance to report noises. This may be because patients do not realize the significance of noises generated by hip joints, or because they may be embarrassed by them. The incidence seems to be significantly higher when patient based surveys are used to determine it. The accepted incidence ranges internationally from <0.5% to 21% [3, 8–12]. These percentages may reflect squeaks and other noises that are often not reported on. These other reported noises are generally excluded from the reported rates. Meticulous interviews should be conducted with patients to determine the exact nature of any reported noises.

Factors associated with squeaks in hip joints may be patient related or may be tied to surgical or implant factors. Certain activities are prone to generate squeaks, these usually being extreme flexion of the hip or cyclical movements. Rising from a low chair, walking and bending are often reported as generating noise. Some studies have also found that squeaks in ceramic-on-ceramic hips are more common in younger, taller and heavier patients. Surgical factors are possibly important, and optimum positioning of the acetabular component may lead to lower squeak rates, but support for this suggestion is not consistent. What is sensible is to accept that malpositioning of components will lead to impingement between the ceramic liner of the acetabular component and the neck of the femoral stem. When revisions have been performed for malpositioning, marks on the edge of the acetabular ceramic component as well as impingement cut marks may be found on the femoral neck. In extreme cases this impingement may lead to chipping and fractures of the ceramic acetabular edge. We have a patient who had a quiet CoC hip until he fell down a flight of stairs. His hip became acutely noisy, with these noises persisting for some time when they were replaced by grinding noises. Radiology revealed fractures of the ceramic liner. Surgery revealed a ceramic liner which had disengaged during the fall, and then re-engaged at an angle. The resultant neck-liner impingement led to circumferential fractures of the ceramic liner (van der Jagt D, Schepers A, 2010). When using hard-on-hard bearings it is becoming increasingly obvious that accurate positioning of the acetabular component is important, and that every effort should be made to position the acetabular component within the range of $40° \pm 10°$ of inclination and $15° \pm 10°$ of anteversion. Not only will this possibly lead to decreased squeak rates, but it should lead to lower wear rates for the bearing [5].

Prosthesis implant factors are well documented, but poorly understood. Because the mechanism of the noise generation is still obscure, various authors have looked at implant designs and their possible contribution to the generation of noises. The multitude of different implant designs, and the different combinations of implants used has confused the subject even further. These design factors can be separated into those on either side of the bearing surface. On the acetabular side, elevated metal rims and locking mechanism at the ceramic insert-metal backing interface have been implicated. On the femoral side, the length and thickness of the femoral neck as well as the materials used in their manufacture may play a role. Relatively more flexible titanium stems may be associated with different squeak rates. Recent work though suggests that in CoC bearings the noise comes from the bearing surface and not from other components. The only way to determine which implant factors are important is to do a meta-analysis using sophisticated statistical methods to isolate the design factors implicated in squeak generation.

Mismatched articular components are particularly prone to squeak. This relates to both the materials used as well as the geometry of the components [13]. Other factors that have been implicated include backside interference in modular components [14] and metal interposition [15]. Loss of lubrication patterns may also be implicated. Finally it should be noted that only CoC and metal-on-metal (MoM) bearings generate squeaks which are sustained [16]. Squeaking noises in MoM bearings are less topical, and seem to occur in both conventional hip replacements as well as those of the resurfacing type. Of interest is that MoM bearings that are of the polyethylene sandwich design do not generate squeaks [16]. This may be because of a dampening effect of the polyethylene.

At the University of the Witwatersrand, Johannesburg, South Africa, we have been conducting an ethics approved, prospective randomized bearing surface trial in an effort to determine the optimum bearing surface. To reduce variable factors to the minimum, all patients received the same femoral component (Corail, DePuy) and metal backed acetabular component (Pinnacle 300, DePuy). All were implanted via a standard antero-lateral approach. Bearing surfaces used were ceramic-on-polyethylene (CoP), MoM, CoC and ceramic-on-metal (CoM). All ceramic components were Biolox delta (Ceramtec/DePuy) and metal components were all cobalt-chrome alloys (Articuleze, DePuy). The polyethylene used was Marathon polyethelene (DePuy). All bearings articulations were 28 mm. This size was settled on firstly so as to not disadvantage the CoP arm of the trial by having a larger higher friction bearing surface, and secondly because larger Biolox delta ceramic components were not yet available. An ongoing clinical and radiological follow-up has been done. Serial whole blood metal ion levels have also been performed [17].

Further ethics approval was obtained to investigate noise emissions in these four patient cohorts [18]. All three hard-on-hard bearing surfaces were investigated, these being the CoC, MoM and CoM cohorts. The CoP cohort of patients served as a control group. As part of the clinical follow-up, detailed interviews were done with all patients, and they were specifically questioned about noises possibly related to their hip replacements. Patients were asked to describe the noises that they had, and also the activity that produced that noise. They were also asked when then noise started, and whether it was still present. Acetabular orientations, as well as whole blood metal ion levels were correlated with the emission of noises.

Our series confirmed a noise incidence of 7% in the CoC group of hips. These were squeaks and excluded other noises such as grinding and clunking. This was consistent with other reported series. The CoP and the MoM cohorts did not squeak, even though there were other noises produced. Of particular significance was the fact that there were no squeaks in the CoM cohort of patients.

In the CoC group, there were no patients with increased whole blood metal ion levels. This led us to conclude that squeak generation in our series was not due to neck-cup impingement, or due to motion at the metal-ceramic morse taper interface, with the resultant fretting leading to elevated whole blood metal ion levels. Furthermore this supports the suggestion by Currier et al. that it is the ceramic components themselves that emit the squeaks [19].

A statistical analysis revealed that the odds of experiencing a noise increased by a factor of 1.96/year. This incremental increase may well mean that noise is related to a deterioration of the articular surface of the bearing. This correlates with the work by van Citters et al. which suggests that squeaking is caused by stripe wear [20]. The roughened surfaces in stripe wear are probably localized surface wear and with time this would be more frequent in incidence in a cohort of bearings, and more extensive over the surface of an individual bearing. This CoC bearing wear could then possibly emit squeaks at an increasing rate correlating with the wear.

Explant retrieval analysis has been performed on CoM bearings and none of these have demonstrated any stripe wear on the ceramic heads [21]. This would further support the suggestion that squeaks in CoC bearings is closely related to stripe wear [17].

A further ethically approved study was done with larger articular bearings [22]. The hips implanted were Corail (DePuy) stems and Pinnacle 300 (DePuy) acetabular metal backed components. In this series, CoC and CoM bearings were investigated, with all bearings being 36 mm in size. A squeak rate of 16% was found in the CoC group. Activities that generated noises included walking and bending. All noisy hips had acetabular components that were well orientated with both inclination and anteversion well within the normal accepted range. Metal ion levels were not recorded in this study. Again of significance, here were no squeaking hips in the CoM cohort.

Even though both surfaces in the CoM bearing couple are hard, the lack of squeaking may be because this hardness is differential. It would be useful to analyze the noises emanating from MoM bearings which are engineered to have a differential hardness between the opposing surfaces. This is achieved through modifying the carbide content differentially between the two halves of the bearing coupling.

In correlating the findings of the two studies, several important conclusions can be made. CoC bearings squeak irrespective of the size of the articulation. Larger 36 mm bearings in our series have a higher squeak rate than smaller 28 mm bearings. Both cohorts had the same cup, stem and locking mechanisms, with the only variable being the size of the bearings and the bearing surfaces. The study further confirmed that CoM bearings did not squeak irrespective of bearing size. One can therefore conclude that the bearing material, and more specifically the CoC, as well as the articular size namely 28 mm compared to 36 mm were contributory to the generation of squeaking noises.

7.1
Summary

Squeaking noises in hips are still poorly understood. Many factors have been implicated in their etiology, but few have been confirmed. Correlation between noise and failure necessitating revision of the hip replacement has also not been shown.

They occur in hard-on-hard bearing surfaces. Ceramic-on-ceramic bearing are particularly prone to squeak. Ceramic-on-metal articulations do not squeak. The bearing size also seems to be important, and with all other factors being equal, 36 mm CoC bearings have a higher squeak rate than 28 mm ones.

References

1. Kuntz, M.: Validation of a new high performance alumina matrix composite for use in total joint replacement. Semin. Arthroplasty **17**, 141–145 (2006)
2. Boutin, P.: Total arthroplasty of the hip by fritted aluminum prosthesis: experimental study and 1st clinical applications. Rev. Chir. Orthop. Reparatrice Appar. Mot. **58**(3), 229–246 (1972)
3. Jarrett, C.A., Ranawat, A.S., Bruzzone, M., Blum, Y.C., Rodriguez, J.A., Ranawat, C.S.: The squeaking hip: a phenomenon of ceramic-on-ceramic total hip arthroplasty. J. Bone Joint Surg. Am. **91**(6), 1344–1349 (2009)
4. Ecker, T.M., Robbins, C., van Flandern, G., Patch, D., Steppacher, S.D., Bierbaum, B., et al.: Squeaking in total hip replacement: no cause for concern. Orthopedics **31**(9), 875–876 (2008). 84
5. Yang, C.C., Kim, R.H., Dennis, D.A.: The squeaking hip: a cause for concern-disagrees. Orthopedics **30**(9), 739–742 (2007)
6. Ranawat, A.S., Ranawat, C.S.: The squeaking hip: a cause for concern-agrees. Orthopedics **30**(9), 738–743 (2007)
7. Rodriguez, J.A., Gonzalez DelaValle, A., McCook, N.: Squeaking in total hip replacement: a cause for concern. Orthopedics **31**(9), 874 (2008). 877–878
8. Keurentjes, J.C., Kuipers, R.M., Wever, D.J., Schreurs, B.W.: High incidence of squeaking in THA's with alumina ceramic-on-ceramic bearings. Clin. Orthop. Relat. Res. **466**(6), 1438–1443 (2008)
9. Walter, W.L., O'Toole, G.C., Walter, W.K., Ellis, A., Zicat, B.A.: Squeaking in ceramic-on-ceramic hips: the importance of acetabular orientation. J. Arthroplasty **22**, 496–503 (2007)
10. Capello, W.N., D'Antonio, J.A., Feinberg, J.R., Manley, M.T., Naughton, M.: Ceramic-on-ceramic total hip arthroplasty: update. J. Arthroplasty **23**(7 suppl), 39–43 (2008)
11. Restrepo, C., Parvizi, J., Kurtz, S.M., Sharkey, P.F., Hozack, W.J., Rothman, R.H.: The noisy ceramic hip: is component malpositioning the cause? J. Arthroplasty **23**(5), 643–649 (2008)
12. Lusty, P.J., Tai, C.C., Sew-Hoy, R.P., Walter, W.L., Walter, W.K., Zicat, B.A.: Third-generation alumina-on-alumina ceramic bearings in cementless total hip arthroplasty. J. Bone Joint Surg. Am. **89**(12), 2676–2683 (2007)
13. Morlock, M., Nassautt, R., Janssen, R., Willmann, G., Honl, M.: Mismatched wear couple zirconium oxide and aluminum oxide in total hip arthroplasty. J. Arthroplasty **16**(8), 1071–1074 (2001)
14. Walter, W.L., Waters, T.S., Gillies, M., et al.: Squeaking hips. J. Bone Joint Surg. Am. **90**(suppl 4), 102–111 (2008)

15. Chevillotte, C., Trousedale, R.T., Chen, Q., Guyen, O., An, K.: The 2009 Frank Stinchfield Award: "Hip squeaking". A biomechanical study of ceramic-on-ceramic bearing surfaces. Clin. Orthop. Relat. Res. **468**, 345–350 (2010)

16. Glaser, D., Komistek, R.D., Cates, H.E., Mahfouz, M.R.: Clicking and squeaking: in vivo correlation of sound and separation for different bearing surfaces. J. Bone Joint Surg. Am. **90**(suppl 4), 112–120 (2008)

17. Williams, S., Schepers, A., Isaac, G., Hardaker, C., Ingham, E., van der Jagt, D., Fisher, J.: The 2006 Otto Aufranc Award: ceramic-on-metal hip arthroplasties: a comparative in vitro an in vivo study. Clin. Orthop. Relat. Res. **465**, 23–32 (2007)

18. Mokete, L., Nwokeyi, K., van der Jagt, D.R., Schepers, A.: Noise emissions in hard-on-hard total hip replacements. In: 54th South African Orthopaedic Association Congress, Cape Town, 1–5 Sept 2008 (2008)

19. Currier, J.H., Reinitz, S.D., Currier, B.H., van Citters, D.W.: Ceramic squeaking: retrieved hips and never implanted components show ceramic components alone emit the squeaking documented in vivo. Paper 161. In: 56th Orthopaedic Research Society Annual Meeting, New Orleans (2010)

20. van Citters, D.W., Currier, J.H., Carlson, E.M., Lyford, K.A., Mayer, M.B.: Imaging of squeaking ceramic surfaces: stripe wear appears likely to drive squeaking more than metal transfer. Paper 2298. In: 56th Orthopaedic Research Society Annual Meeting, New Orleans (2010)

21. Isaac, G.H., Brockett, C., Breckon, A., van der Jagt, D., Williams, S., Hardaker, C., Fisher, J., Schepers, A.: Ceramic-on-metal bearings in total hip replacement. Whole blood metal ion levels and analysis of retrieved components. J. Bone Joint Surg. Br. **91-B**, 1134–1141 (2009)

22. Gelbart, B., Mokete, L., van der Jagt, D.R., Nwokeyi, K., Schepers, A.: Noise emissions with hard-on-hard bearings in total hip replacements – 28 mm heads compared to 36 mm heads. In: 20th Biennial South African Arthroplasty Congress, Sun City, 18–22 Mar 2009 (2009)

Head Size in Relation to Noise Occurrence in Ceramic-on-Ceramic Bearings

8

Frank Hoffmann, Milan Jovanovic, and Michael Muschik

8.1
Introduction

Ceramic-on-ceramic (COC) has been an excellent alternative bearing surface for total hip arthroplasties (THA) in young, high-demand patients with end-stage arthritis of the hip [1, 2]. Ceramic material has been used for THA in Europe for 40 years with variable results [2–4]. Hamadouche et al. described minimal wear, a low rate of complications and limited osteolysis with COC THA after 18.5 years of follow-up [1]. On the other hand, the revision rate in Europe and the USA from 1988 till 1996 varied between 3% and 44% [5].

With ceramic on ceramic bearing surfaces in total hip arthroplasty, audible noise is a frequently documented problem. The frequency of squeak has been reported between 0.7% and 20.9% [6, 7].

Bigger heads should theoretically have more range of motion (ROM) (Fig. 8.1).

With increasing head size, there is less chance of component-on-component impingement and therefore a decreased risk of (sub)luxation chipping, breaking and maybe noise occurrence (Table 8.1).

8.2
Material and Methods

We performed a prospective multicenter study on 152 patients (92 males and 60 females), with the non-cemented modular pressfit cup seleXys (Mathys Ltd Bettlach, Switzerland).

F. Hoffmann (✉)
Clinic for Orthopaedic Surgery and Sports Traumatology, RoMed Klinikum Rosenheim,
Pettenkoferstr. 10, Rosenheim D-83022, Germany
e-mail: frank.hoffmann@ro-med.de

M. Jovanovic
Orthopädisches Zentrum Münsingen, Münsingen, Switzerland

M. Muschik
Park-Klinik Weissensee, Berlin, Germany

K. Knahr (ed.), *Tribology in Total Hip Arthroplasty*,
DOI: 10.1007/978-3-642-19429-0_8, © 2011 EFORT

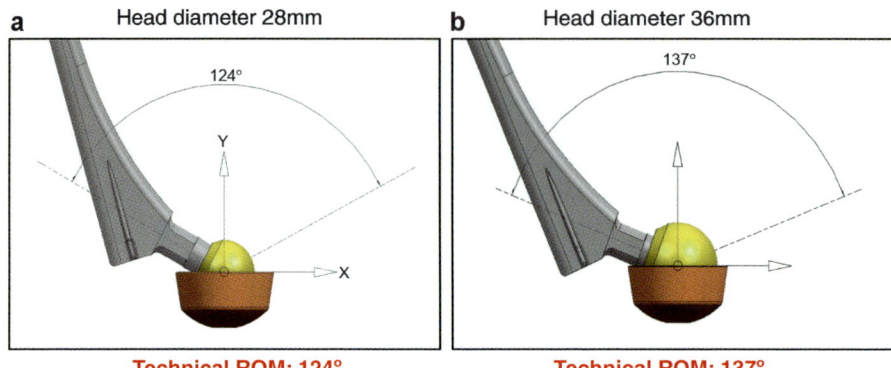

Fig. 8.1 (a) Technical ROM: 124°. (b) Technical ROM: 137°

Table 8.1 Range of flexion dependent on the anteversion of the cup

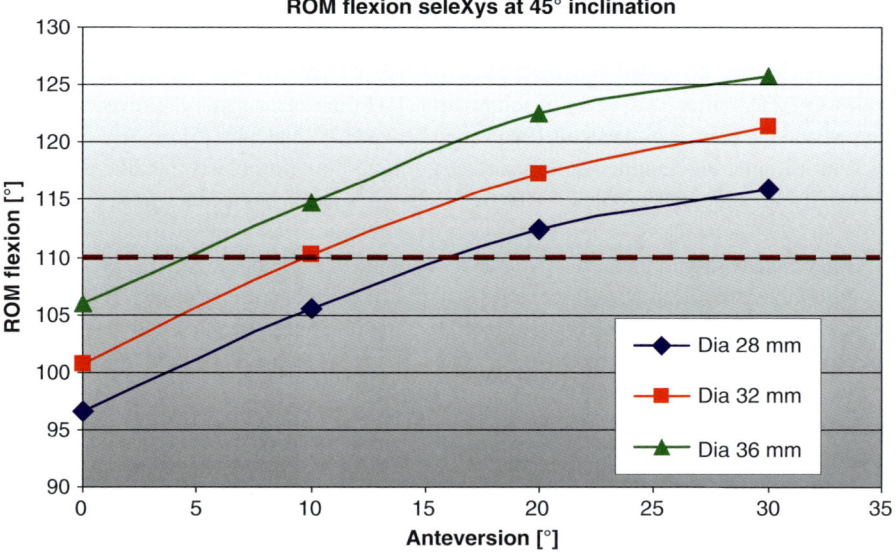

There were seven patients with bilateral surgery. 113 hips were implanted minimally invasive, 37 conventional.

All hips had a ceramic on ceramic articulation with a femoral head size of 32 mm or 36 mm (Fig. 8.2).

The ceramic consists of 100% Al,O, with a grainsize of 2.3 μm.

The mean age at surgery was 67.1 years (females 69.1 years, males 65.9 years). The BMI was 27.9 kg/m² on average (females 27.7 kg/m², males 28.1 kg/m²).

The indication for surgery was osteoarthritis in 89.5%, avascular necrosis in 5.9% and others in 4.6%.

Preoperatively we collected the Harris Hip Score and the VAS for pain and satisfaction and we did an X-ray evaluation.

The clinical controls were done at 3, 6, 12 and 24 months postoperatively, again with X-ray evaluation, collection of the Harris Hip Score and the VAS for pain and satisfaction.

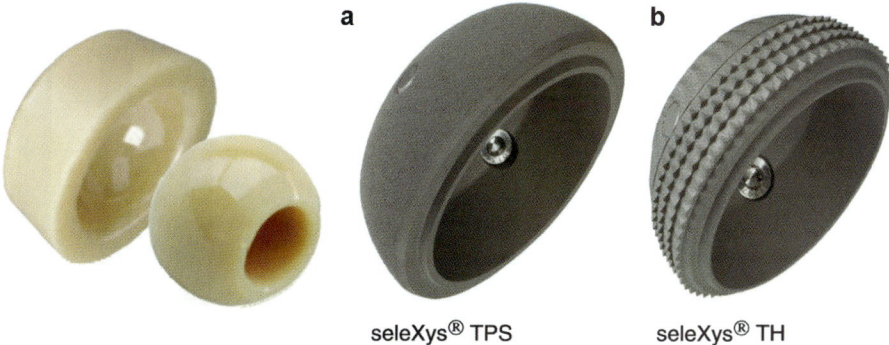

Fig. 8.2 (a) seleXys® TPS. (b) seleXys® TH

8.3
Results

Patients with femoral heads of 32 and 36 mm diameter had excellent and comparable clinical results. In the group with 32 mm heads we found 79% females whereas in the 36 mm group, only 16% were female.

The Harris Hip Score increased from a preoperative mean of 49–96 points (Table 8.2).

Table 8.2 Harris Hip Score over time

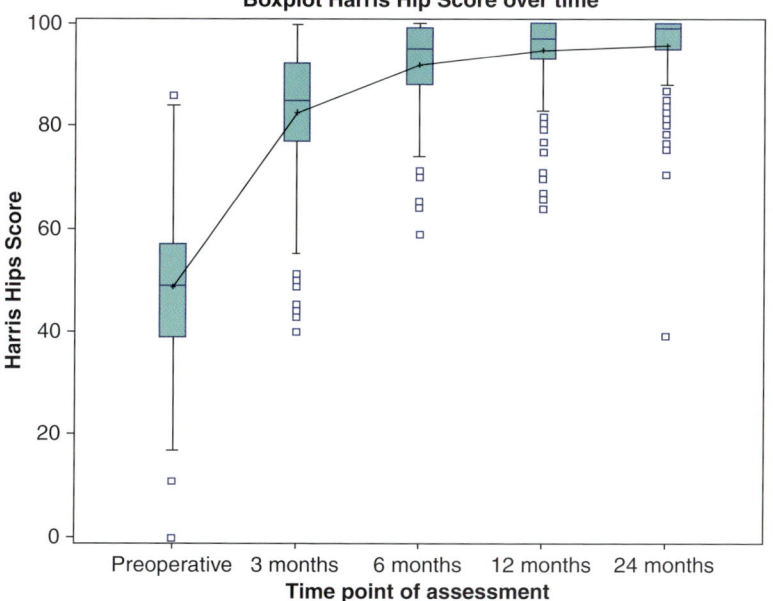

Table 8.3 VAS pain and satisfaction over time

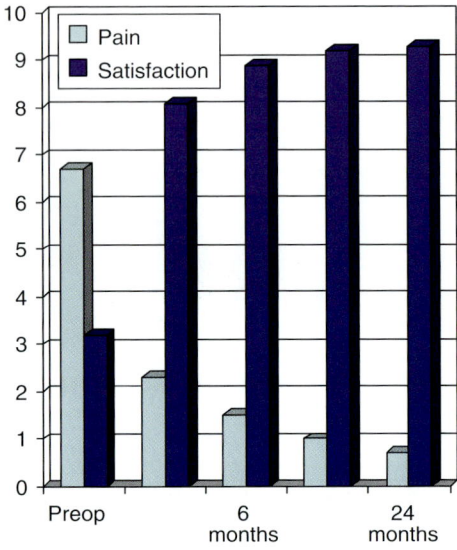

Table 8.4 ROM in degrees pre- and postoperatively

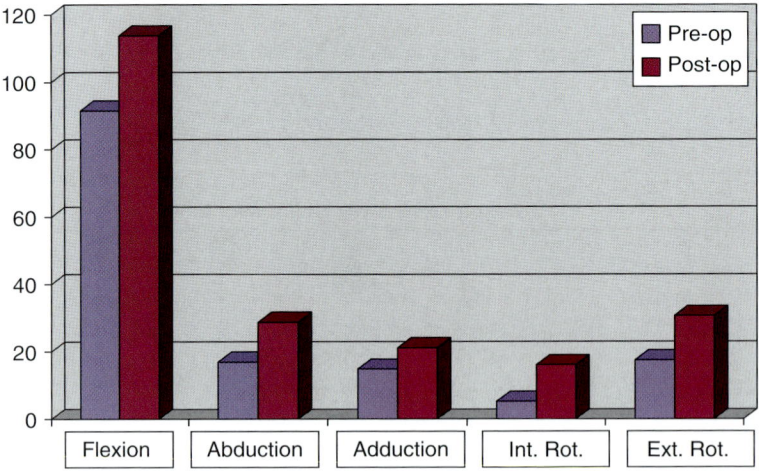

The VAS for pain improved from 6.7 to 0.8, and for satisfaction from 3.1 to 9.2 (Table 8.3).

The range of motion (ROM) for flexion increased from a mean of 91.6–113.8°, for abduction from 17.0° to 28.8°, for adduction from 14.9° to 21.2°, for internal rotation from 5.3° to 16.1° and for external rotation from 17.5° to 30.6° (Table 8.4).

As intraoperative complications, one femur fissure and two trochanter major fractures were noted. The postoperative complications were one wound revision due to a deep infection, five luxations and one squeaking hip.

This patient (male, 74 years, weight 79 kg, height 165 cm) had a squeaking p.o., but no pain. Six months later he had a feeling of subluxations. After 12 months a cracking was audible.

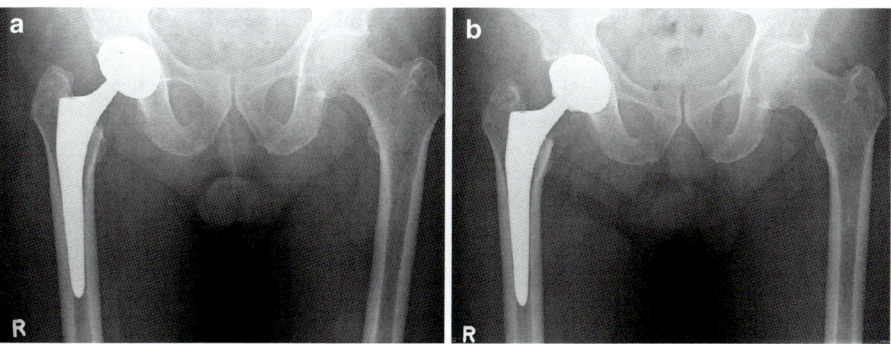

Fig. 8.3 (**a**) Before revision. (**b**) After revision

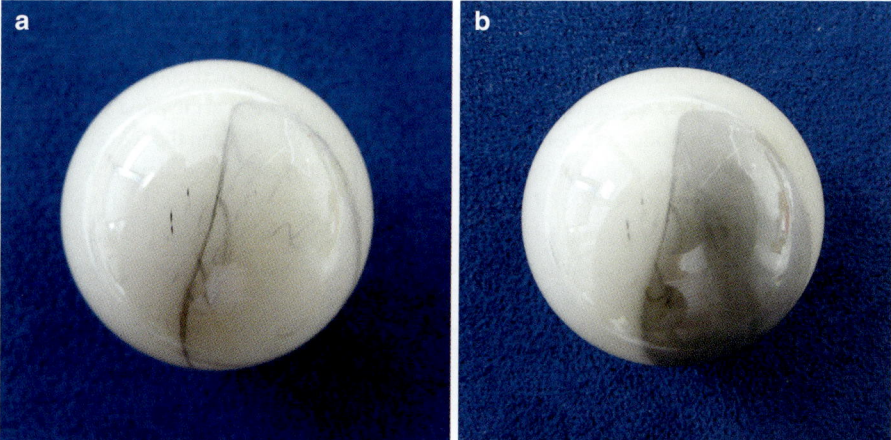

Fig. 8.4 (**a**) Stripe wear in the polar area of the ceramic head (marked with a *pencil*). (**b**) Stripe wear in the polar area of the ceramic head (marked with *graphite*)

A revision from a 36 to a 32 mm head and change of the ceramic liner in a PE insert was performed after 13 months. Afterwards the squeaking was resolved (Fig. 8.3a, b). The removed ceramic head showed significant stripe wear (Fig. 8.4a, b)

8.4
Discussion

Due to minimal wear, ceramic-on-ceramic (COC) has been an excellent alternative bearing surface for THA in young, high-demand patients. But recently described hip noise (squeak) associated with COC bearings has caused worry among physicians and patients [7, 8].

In an Australian study, the incidence of squeaking in COC hips was 0.7% [7]. They reported a higher variance in acetabular anteversion and inclination in the hips that squeaked.

A study from The Netherlands reported a 20.9% incidence of squeaking in 43 noncemented COC hips [6]. The study showed no difference in patient characteristics or acetabular placement between squeaking and non squeaking hips, but found short necks on the implants in hips that squeaked.

Mai et al. from the USA had 10% squeaking hips from 336 COC THA [9]. Most of the squeaking initially occurred between 12 and 30 months postoperatively but could occur also as late as 4 years postoperatively. Most squeakers (29 of 32 THAs; 91%) stated they were pain-free and doing well despite the occasional noise. Patients who experienced squeaking were taller than non squeaking patients. Squeaking occurred in a higher percentage of patients with femoral components having smaller neck geometry.

Taylor et al. produced wear stripes on ceramic bearings in a laboratory setting and determined that under certain conditions, noise could occur during either edge loading or normal articulation [10]. Other mechanisms that have been postulated are the violation of a lubrication layer or component impingement.

Chevillotte et al. found in an in vitro bench test that squeaking occurred in all situations without lubrication (normal gait, high load, stripe wear, material transfer, edge wear, microfractures) [11]. In situations with lubrication, squeaking was generated only when material transfer occurred. They suggested that squeaking noise in COC bearings is a problem of COC lubrication. This phenomenon occurs when the fluid film between the two surfaces in contact is disrupted. This interruption is most commonly the result of metal particle transfer as a third body between COC. Metal transfer as a primary mode leading to fluid lubrication disruption may explain why squeaking is more common in certain protheses designs.

We could show in our study that an increased range of motion is no cause of squeaking if one uses large ceramic heads with 32 and 36 mm diameter. We had only one patient who developed a squeaking noise. In this case, the cup had a steep anteversion with the possible dorsal neck impingement.

8.5
Conclusion

The non-cemented modular cup combined with a ceramic inlay and a ceramic femoral head is a safe implant with good clinical results after 2 years. All patients undergoing total hip arthroplasty showed significant improvements in postoperative functioning and activity level after the implantation of a selcXys cup.

References

1. Hannouche, D., et al.: Ceramics in total hip replacement. Clin. Orthop. Relat. Res. **430**, 62–71 (2005)
2. Sedel, L.: Evolution of aluminia-on-aluminia implants: a review. Clin. Orthop. Relat. Res. **379**, 48–54 (2000)

3. Boutin, P.: Arthroplastie totale de la hanche par prothese en alumine frittee. Rev. Chir. Orthop. Reparatrice Appar. Mot. **58**(3), 229–246 (1972)
4. Mittelmeir, H.: Selbsthaftende Keramik-Metall-Verbund-Endoprothesen. Med. Orthop. Tech. **95**(6), 152–159 (1975)
5. Clarke, I.C., et al.: Tribological and material properties for all-aluminia THR – Convergence with clinical retrieval data. In: Zippel, H., Dietrich, M. (eds.) Bioceramics in Joint Arthroplasty. Steinkopff, Darmstadt (2003)
6. Keurentjes, J.C., Kuipers, R.M., Wever, D.J., et al.: High incidence of squeaking in THAs with aluminia ceramic-on-ceramic bearings. Clin. Orthop. Relat. Res. **466**, 1438–1443 (2008)
7. Wl, W., et al.: Squeaking in ceramic-on-ceramic hips: the importance of acetabular component orientation. J. Arthroplasty **22**, 496–503 (2007)
8. Rodriguez, J.A., et al.: Squeaking in total hip replacement: a cause of concern. Orthopaedics **31**(874), 877–878 (2008)
9. Mai, K., et al.: Incidence of "squeaking" after ceramic-on-ceramic total hip arthroplasty. Clin. Orthop. Relat. Res. **468**, 413–417 (2010)
10. Taylor, S., et al.: The role of stripe wear in causing acoustic emissions from aluminia ceramic-on-ceramic bearings. J. Arthroplasty **22**(Suppl 3), 47–51 (2007)
11. Chevillotte, C., et al.: "Hip squeaking". A biomechanical study of ceramic-on-ceramic bearing surfaces. Clin. Orthop. Relat. Res. **468**, 345–350 (2010)

The Squeaking Phenomenon in Ceramic-on-Ceramic Bearings

9

Alexandra Pokorny and Karl Knahr

9.1
Introduction

The growing demand for total hip arthroplasties in an increasingly younger, more active group of patients has led to new challenges for tribological properties of implants.

Aiming at optimizing wear rates, ceramic-on-ceramic bearings have become popular. They feature excellent biocompatibility and superb clinical results with a reported survival rate of more than 85% in cementless fixation at long-term follow-up of a mean 19.7 years [11]. Intermediate-term outcome can even reach a survival rate of 96% [15].

Recent reports on audible sensations however are causing wide concern. The incidence of a distinct squeaking varies between 0.3% and 10.7% [14, 15]. Various other noises like clicking or grinding can reach an incidence of 32.8% [14].

The etiology remains unclear and is most likely multifactorial. Theories include prosthetic design, malpositioning of components, edge loading, stripe wear and patient activity levels.

9.2
Historical Development of Ceramic Bearings

Survivorship of total hip arthroplasty is largely affected by wear of the bearing components leading to polyethylene or metal debris, which in turn can trigger osteolysis. Ceramic-on ceramic (C-o-C) bearings show the lowest wear rates [24] and outstanding biocompatibility and do not seem to trigger excessive immune reactions. The first generation of ceramic bearings in Europe was introduced by Boutin in France in 1972 [2, 3], followed by Mittelmeier and Griss in Germany [10, 17] and Salzer in Austria in 1976 [20, 21].

A. Pokorny (✉) and K. Knahr
Orthopaedic Hospital Vienna-Speising, 2nd Orthopaedic Department,
Speisinger Straße 109, 1130 Vienna, Austria
e-mail: alexandra.pokorny@oss.at

K. Knahr (ed.), *Tribology in Total Hip Arthroplasty*,
DOI: 10.1007/978-3-642-19429-0_9, © 2011 EFORT

9

They were expected to solve problems like wear rate, biocompatibility, cementless fixation by good bone ongrowth and most importantly osteolysis associated with polyethylene and metal bearings. The early results were encouraging; nevertheless there was a high incidence of failure due to component fracture and loosening. Large grain size as well as an adverse broad distribution of grain size and an inferior quality of alumina were the underlying causes.

Subsequently new generations of C-o-C bearings were developed eliminating this risk and presenting the superior tribological properties of ceramics. The introduction of hot isostatic pressing led to the breakthrough. Nowadays the risk for head fracture ranges between 0.004% and 1.4% and the risk for liner fracture is reported to be as low as 0.01–2% [12].

The in vitro wear rate of ceramics by far outstands other bearings – especially the widely used metal-on-polyethylene. In 2003, the FDA granted approval for the use of third-generation ceramic bearings in the USA.

9.3
Incidence of Squeaking

Large interest was sparked in 2006, when various publications reported a squeaking sensation emanating from C-o-C bearings [15, 26, 27]. The incidences varied widely, ranging from 0.3% to 10.7% [14, 15] (Table 9.1).

Table 9.1 Overview of incidences of squeaking

Incidence of squeaking in ceramic-on-ceramic bearings		
0.3%	1/301	Secur-Fit or Secur-Fit plus+Osteonics ABC cup Lusty et al. [15]
0.7%	17/2397	ABG II system, Osteonics THR, ABG II stem+Duraloc option cup, various combinations Walter et al. [29]
0.8%	3/452	Omnifit stem+4 groups: ABC insert, porous coated shell, arc deposited HA coated shell, Trident cup Capello et al. [30]
2.7%	30/1056	Omnifit or Accolade stem+Trident cup Restrepo et al. [31]
10.3%	18/175	ABG II stem+cup Grimm et al. [32]
10.7%	14/131	Accolade TMZF stem+Trident PSC cup Jarret et al. [14]

9.4
Influencing Factors

Numerous factors have been reported as the cause for this phenomenon. Generally, all publications agree that the etiology is multifactorial.

9.4.1
Prosthetic Design and Material

Recent reports emphasize the importance of prosthetic design and material. Squeaking is caused by oscillations of the implant due to vibrations. Experimental analysis by Weiss et al. [28] has shown that these vibrations are generated by an instability of the relative motion of the components with respect to each other. Some stems were found to be more susceptible than others.

But also in clinical analysis, some implant design and materials seem to favour the spreading of vibrations. Restrepo et al. [19] demonstrated the importance of the femoral stem component. He found an increased incidence of squeaking in patients managed with a thinner stem component with a V-40 taper neck and a stem made of titanium-molybdenum-zirconium-iron alloy, thus indicating that the femoral stem plays a vital role in the development of acoustic phenomena.

Furthermore, the design of the cup seems to play a key role. Numerous reports on acoustic emissions from THAs with a cup design consisting of a metal rim on a ceramic liner can be found. The possible impingement between the metal rim and the neck of the prosthesis, especially in designs with a short neck, might lead to increased edge loading or wear of the bearings or even to third body wear, which in turn can damage the surface of any bearing and cause stripe wear. Swanson et al. [23] found a high incidence of squeaking of up to 35.6% for those ensleeved designs. 11.1% of patients experienced an audible sensation at least once a week. Short femoral neck length seemed to favour the occurrence.

9.4.2
Fluid Film Disruption

Chevilotte et al. [5] conducted in vitro testing of 32 mm heads and inserts in varying conditions. Interestingly, squeaking was reproducible in all dry conditions, especially with increased loading, stripe wear and metal transfer. Furthermore, in case of material transfer, squeaking could be even found in lubricated conditions, whereas in the other settings squeaking disappeared when lubrication was introduced. Therefore, the authors concluded that a disruption of the lubrication layer is the underlying cause for the squeaking phenomenon.

9.4.3
Microseparation

Another closely linked factor to these findings seems to be microseparation of the components. Nevelos et al. [18] and Glaser et al. [9] demonstrated microseparation for all bearings during normal gait. This seems to be aggravated by a general joint laxity. Microseparation in turn leads to edge loading and increased wear of the components. The so called stripe wear is formed. In an area of such wear, increased amounts of friction might be generated, thus leading to vibration.

9.4.4
Component Orientation

Initial reports on squeaking attribute a major role to anteversion and inclination of the acetabular component [26, 27]. W. Walter evaluated 2,716 THAs with ceramic on ceramic bearings and found an increased variance in cup anteversion for squeaking hips leading to a recommendation for positioning of the acetabular component within 10° of a target of 25° operative anteversion and operative inclination of 45°. Outside this range, he determined an increased possibility of 29% for squeaking. In contrast, Restrepo et al. [19] could not verify a correlation between cup anteversion or inclination and the occurrence of squeaking. This finding is supported by various other reports [22].

9.4.5
Patient Related Factors

Walter et al. [27] described a correlation between patient age, BMI and activity level and the occurrence of squeaking, which could be found more frequently in the overweight, young, active patient.

9.5
Clinical Experience

Until increasing literature coverage of the squeaking sensation we were unaware of patient reports of acoustic phenomena. Therefore we started a retrospective analysis of our ceramic-on-ceramic total hip arthroplasties in 2008.

9.5.1
Materials and Method

To investigate the occurrence of acoustic emissions in our patients, we conducted a retrospective study of a consecutive series of patients, who had all received the same prosthesis system (Zimmer™ Alloclassic Variall®) in combination with ceramic-on ceramic bearings. The aim was to evaluate the nature of the noise, duration and clinical consequence.

First introduction of the Alloclassic® hip system was conducted in 1979. After some modifications, the Alloclassic Variall ® system emerged and has been in use at our department since 1998. This cementless design features a rectangular titanium stem of Protasul titanium alloy TiAlNb, which creates a diaphyseal press fit. Secondary stability is achieved by bone ongrowth onto the grist-blasted surface (Fig. 9.1).

The conical acetabular component is threaded into the bone and gains long-term stability by bone ongrowth on its rough titanium surface.

This hip system was combined with 28 mm heads of alumina ceramic-on-ceramic Cerasul® bearing available by Zimmer™. Cerasul® is a third-generation aluminum oxide (Al_2O_3) hot isostatic pressed ceramic, first implanted in Europe in 1998. It is available in three head diameters: small, medium, large.

The alumina Cerasul® gamma inlays were used. To date no fracture of Cerasul® Gamma inlays has been reported to Zimmer™. The fracture rate for the Cerasul® heads ranges around 1:6,2000 (0.01%).

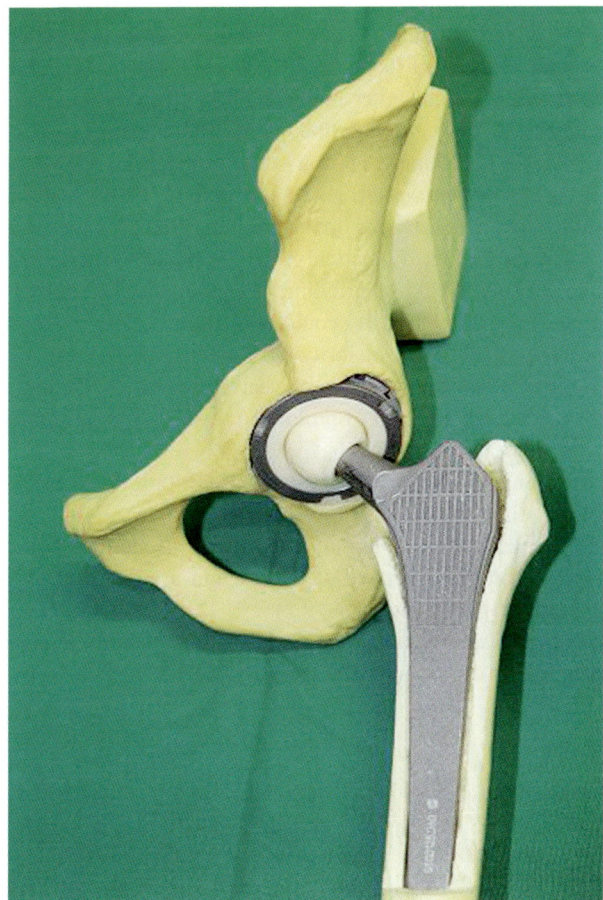

Fig. 9.1 Zimmer™
Alloclassic Variall® system

Between 1998 and 2003, a consecutive series of 327 patients received 337 cementless Zimmer™ Alloclassic Variall implants at our department. This secured a minimum follow-up period of 5 years.

Patients were operated by different surgeons of the same department using the standard lateral transgluteal (Bauer) approach.

In order to conduct a retrospective analysis of the occurrence of audible sensations, patients received a detailed questionnaire via mail, including questions on the first occurrence of squeaking, information on the kind of noise, duration of the phenomenon and possible negative subjective evaluation on behalf of the patient. In case of a positive reply, the patient was invited to a clinical exam and radiographic evaluation. In addition, a specialised audiography was conducted in patients, who reported audible sensations. Occurrence of noise was tested walking, bending and on clinical exam.

9.5.2
Results

Two hundred and twenty-nine patients returned the questionnaire, 21 were deceased and 46 could not be contacted due to change of address or refused to participate in this study.

Only one patient (0.4%), a 52 year old female, reported a distinct squeaking, which first occurred 98 months after implantation. Initially it was not associated with pain, but soon aggravated. However, only the questionnaire sent by mail caused her to seek contact. She reported a squeaking occurring with every movement, which did not keep her from being physical active but was perceived as disturbing. On clinical exam, movement was painful. The subsequent X-ray showed implant failure with pelvic protrusion of the acetabular implant.

Thirty-one (13.5%) patients reported to have experienced varying other types of noises like clicking, creaking, grinding or combinations of noises. In three cases (9.7%) a snapping of the iliotibial band could be identified. The mean onset of noise was 45.6 months postoperatively. The majority of patients (83.9%) experienced the noise with specific movement like getting up from a seated position or bending. Four patients (13.9%) reported the onset after a period of prolonged walking and one patient (3.2%) felt a clicking noise with every movement.

In some cases (16%) noise was self limiting. Other patients could avoid the occurrence by adapting specific positions (29%).

In order to validate impingement as a possible cause we evaluated the neck length of the component. However we could not find a significant difference between hips emitting noises and silent hips (Table 9.2).

Demographic analysis showed no significant difference in gender, age or BMI between patients with noisy and silent hips (Table 9.3).

9.5.3
Revisions

More than half of the patients (52%) who reported audible sensations, felt disturbed by the noise, with one patient seeking revision surgery for this cause. She felt increasingly limited by clicking noises emanating from the hip implant.

Table 9.2 Comparison of neck length for noisy and silent hips

Neck length	Noisy hips (%)	Silent hips (%)
Short	16	6
Medium	23	35
Large	61	57

Table 9.3 Comparison of patient factors for noisy and silent hips

		Noisy hips	Silent hips
Gender	Male (%)	39	32
	Female (%)	61	68
Weight (kg)		77.83±19.2	78.5±15
Height (cm)		169.5±11.9	167±9
BMI		29.7±3.7	28.1±4.4

In addition, two further patients required revision surgery. One 56 year old male patient reported a creaking sensation, which started 108 months postoperatively and was later associated with pain. Another 53 year old female patient perceived a clicking sensation associated with pain. Ceramic components were analysed by Ceramtec™. Analysis of the retrievals showed areas of increased wear corresponding to episodes of edge loading as occurring in subluxation. In addition, a distinct area of metal transfer could be found on one of the ceramic inserts corresponding to impingement (Fig. 9.2–9.4).

Therefore we found a total rate of revision for noise of 1.7% (four patients).

Fig. 9.2 Metal transfer on ceramic liner

9

Fig. 9.3 Area of stripe wear
on ceramic liner

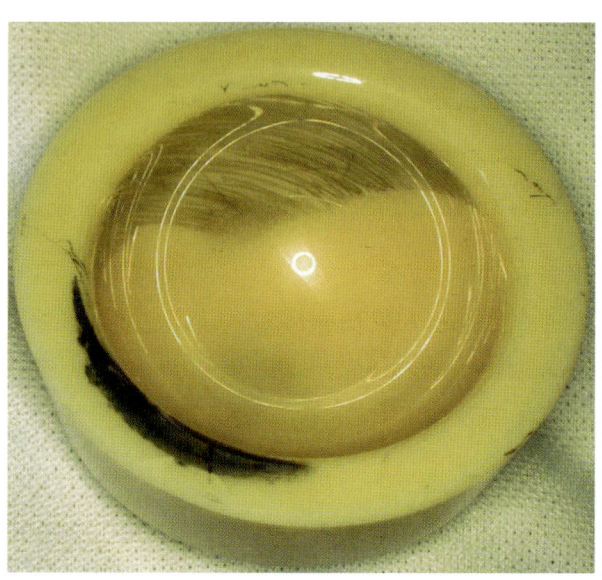

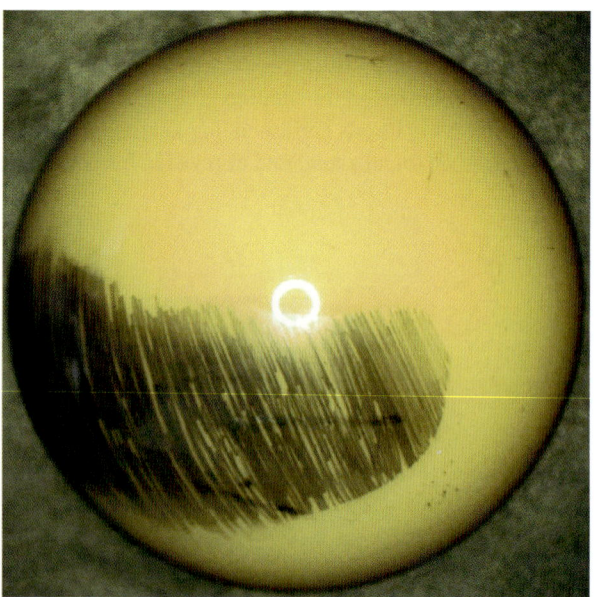

Fig. 9.4 Area of stripe wear
on ceramic head

9.6
Discussion

Studies, which describe high incidences of squeaking or other acoustic phenomena, often deal with the same specific implant groups [14, 15, 27], which seem to favour the development of acoustic emissions because of their design. A characteristic of the cementless Alloclassic Variall® system is the thick metal threaded shell acting like a tight screw in the bone as well as the tapered stem with standard length. This might prevent the spreading of vibrations generated by the bearing couples. Furthermore the rather long and narrow neck of the Variall® stem might lower the risk of impingement.

Overall, we could not find a correlation between neck size and occurrence of audible sensations.

During revisions however, we intraoperatively found increased joint play leading to possible subluxation in three of four revision cases. Subluxation in turn might lead to increased isolated wear of the components creating stripe wear and possibly third wear particles due to direct contact with the metal shell.

Generally, audible sensations are not limited to C-o-C bearings [1, 4, 6, 7, 13]. They were first mentioned as soon as the 1950s for the Judet acrylic hemiarthroplasty. Later on Holzmann [13] reported a clicking noise for Metal-on-Metal bearings for 18 of 117 hips. A transient squeaking was described by Brockett et al. [4] for large diameter Metal-on-Metal bearings as well as by Back et al. [1] in a Metal-on-Metal resurfacing hip.

Recent publications by Glaser et al. [9] and Clarke et al. [6] indicate that all kinds of bearings can cause acoustic sensations. Further reports on the emission of noises for metal-on-metal bearings and hip resurfacings [8] support these findings.

In general, "noisy hips" are more frequent than expected, though they hardly require intervention. But noise can be the first sign of an underlying serious problem like implant failure, malpositioning, impingement and advanced wear and can adversely affect patient satisfaction.

Toni et al. [25] reported the occurrence of noise as an early clinical sign of liner chipping or fracture or stripe wear of the head. This is in line with our findings of a case of squeaking in the presence of failure of the acetabular component.

We found a very low incidence of audible sensations for ceramic bearings in combination with the Alloclassic Variall® system. Therefore we conclude that the generation of noise seems to be affected by choice of prosthetic design. High incidences linked to other implant systems are supporting this assumption and show in vivo that the generation of noises is a complex interaction between bearing and implant.

The development of the new delta ceramic might be a possible solution to the squeaking phenomenon. The delta ceramic consists of a combination of 82% alumina and 17% zirconia as well as 0.5% chromium oxide and 0.3% strontium. This combination improves wear characteristics and makes the bearing less prone to fracture by diffusing crack energy. One recent report by Hamilton et al. [12] of a prospective, randomized, multicenter trial of 263 patients (264 hips) could find no squeaking for this type of ceramic bearing.

9

Should revision of a ceramic-on-ceramic implant due to squeaking become necessary, recent data has shown that a change to metal-on-highly crosslinked polyethylene eliminates squeaking and shows promising results [16]. However long term results are outstanding and possible adverse effects due to change from a hard-on-hard to a hard-on-soft bearing could occur.

In any case, safe primary handling of the rather delicate components when using ceramic-on-ceramic bearings is the key to a satisfactory outcome.

References

1. Back, D.L., Dalziel, R., Young, D., Shimmin, A.: Early results of primary Birmingham hip resurfacings. An independent prospective study of the first 230 hips. J. Bone Joint Surg. Br. **87**(3), 324–329 (2005)
2. Boehler, M., Knahr, K., Plenk Jr., H., Walter, A., Salzer, M., Schreiber, V.: Long-term results of Uncemented Alumina Acetabular implants. J. Bone Joint Surg. Br. **76-B**, 53–59 (1994)
3. Boutin, P.: Total arthroplasty of the hip by fritted aluminium prosthesis. Experimental study and 1st clinical applications. Rev. Chir. Orthop. Reparatrice Appar. Mot. **58**, 29–46 (1972); French
4. Brockett, C.L., et al.: The influence of clearance on friction, lubrication and squeaking in large diameter metal- on- metal hip replacements. J. Mater. Sci. Mater. Med. **19**(4), 1575–1579 (2008)
5. Chevillotte, C., Trousdale, R.T., Chen, Q., Guyen, O., An, K.: The 2009 Frank Stinchfield Award. "Hip squeaking". A biomechanical study of ceramic-on-ceramic bearing surfaces. Clin. Orthop. Relat. Res. **468**, 345–350 (2010)
6. Clarke, I., Manley, M.T.: How do alternate bearing surfaces influence wear behavior? J. Am. Acad. Orthop. Surg. **16**(Suppl 1), S86–S93 (2008)
7. Dumbleton, J.H., Manley, M.T.: Metal-on-metal total hip replacement. What does the literature say? J. Arthroplasty **20**(2), 174–188 (2005)
8. Esposito, C., Walter, W.L., Campbell, P., Roques, A.: Squeaking in metal-on-metal hip resurfacing arthroplasties. Clin. Orthop. Relat. Res. **468**, 2333–2339 (2010)
9. Glaser, D., Komistek, R.D., Cates, H.E., Mahfouz, M.R.: Clicking and squeaking: in vivo correlation of sound and separation for different bearing surfaces. J. Bone Joint Surg. Am. **90**, 112–120 (2008)
10. Griss, P., Heimke, G.: Five years' experience with ceramic-metal-composite hip endoprostheses. I. Clinical evaluation. Arch. Orthop. Trauma. Surg. **98**, 157–164 (1981)
11. Hamadouche, M., Boutin, P., Daussange, J., Bolander, M.E., Sedel, L.: Alumina-on-alumina total hip arthroplasty: a minimum 18.5 year follow up study. J. Bone Joint Surg. Am. **84**, 69 (2002)
12. Hamilton, W., McAuley, J., Dennis, D., Blumenfeld, T., Politi, J.: THA with delta ceramic. Clin. Orthop. Relat. Res. **468**, 358–366 (2010)
13. Holzmann, P., Eggli, S., Ganz, R.: Metal-on-metal: all things bright and beautiful in opposition. Orthopedics **25**, 932 (2002)
14. Jarrett, C.A., Ranawat, A.S., Bruzzone, M., Blum, Y.C., Rodriguez, J.A., Ranawat, C.S.: The squeaking hip: a phenomenon of ceramic-on-ceramic total hip arthroplasty. J Bone Joint Surg. Am. **91**, 1344–1349 (2009)
15. Lusty, P.J., Tai, C.C., Sew-Hoy, R.P., Walter, W.L., Walter, W.K., Zicat, B.A.: Third generation alumina-on-alumina ceramic bearings in cementless total hip arthroplasty. J. Bone Joint Surg. Am. **89**, 2676–2683 (2007)

16. Matar, W.Y., Restrepo, C., Javad, P., Kurtz, S.M., Hozack, W.J.: Revision hip arthroplasty for ceramic-on-ceramic squeaking hips does not compromise the results. J. Arthroplasty **25** (6 Suppl), 81–86 (2010). Epub 15 Jul 2010

17. Mittelmeier, H.: Report on the first decennium of clinical experience with a cementless ceramic total hip replacement. Acta Orthop. Belg. **51**, 367–376 (1985)

18. Nevelos, J., Ingham, E., Doyle, C., Streicher, R., Nevelos, A., Walter, W., Fisher, J.: Microseparation of the centers of alumina-alumina artificial hip joints during simulator testing produces clinically relevant wear rates and patterns. J. Arthroplasty **15**, 793–795 (2000)

19. Restrepo, C., Post, Z., Kai, B., Hozack, W.: The effect of stem design on the prevalence of squeaking following ceramic-on-ceramic bearing total hip arthroplasty. J. Bone Joint Surg. Am. **92**, 550–557 (2010)

20. Salzer, M., Knahr, K., Locke, H., et al.: A bioceramic endoprosthesis for the replacement of the proximal humerus. Arch. Orthop. Trauma. Surg. **93**, 169–184 (1979)

21. Salzer, M., Zweymüller, K., Locke, H., et al.: Further experimental and clinical experience with aluminium oxide endoprostheses. J. Biomed. Mater. Res. **10**, 847–856 (1976)

22. Schroder, D., Bornstein, L., Bostrom, M.P.G., Nestor, B.J., Padgett, D.E., Westrich, G.H.: Ceramic-on-ceramic total hip arthroplasty. Incidence of instability and noise. Clin. Orthop. Relat. Res. **469**(2), 437–442 (2011). Epub 18 Sep 2010

23. Swanson, T., Peterson, D.J., Seethala, R., Bliss, R., Spellmon Jr., C.A.: Influence of prosthetic design on squeaking after ceramic-on-ceramic total hip arthroplasty. J. Arthroplasty **25** (6 Suppl), 36–42 (2010)

24. Tipper, J.L., Firkins, P.J., Besong, A.A., Barbour, P.S.M., Nevelos, J., Stone, M.H., Ingham, E., Fisher, J.: Characterisation of wear debris from UHMWPE on zirconia ceramic, metal-on-metal and alumina ceramic-on-ceramic hip prostheses generated in a physiological anatomical hip joint simulator. Wear **250**, 120–128 (2001)

25. Toni, A., Traina, F., Stea, S., Sudanese, A., Visentin, M., Bordini, B., Squarzoni, S.: Early diagnosis of ceramic liner fracture. Guidelines based on a twelve-year clinical experience. J. Bone Joint Surg. Am. **88**(Suppl 4), 55–63 (2006)

26. Walter, W., Insley, G.M., Walter, W.K., Tuke, M.A.: Edge loading in third generation alumina ceramic-on-ceramic bearings. J. Arthroplasty **19**, 402–413 (2004)

27. Walter, W., Waters, T.S., Gillies, M., Donohoo, S., Kurtz, S.M., Ranawat, A.S., Hozack, W.J., Tuke, M.A.: Squeaking hips. J. Bone Joint Surg. Am. **90**, 102–111 (2008)

28. Weiss, C., Gdaniec, P., Hoffmann, N., Hothan, A., Huber, G., Morlock, M.: Squeak in hip endoprosthesis systems: An experimental study and a numerical technique to analyse design variants. Med. Eng. Phys. **32**(6), 604–609 (2010). Epub 16 Mar 2010

29. Walter, W., et al.: Squeaking in ceramic-on-ceramic hips: The importance of acetabular component orientation. J. Arthroplasty **22**(4), 496–503 (2007)

30. Capello, W.N., D'Antonio, J.A., Feinberg, J.R., et al.: Ceramic-on-ceramic total hip arthroplasty: update. J. Arthroplasty **23**(7 Suppl), 39–43 (2008)

31. Restrepo, C., Parvizi, J., Kurtz, S.M., Sharkey, P.F., Hozack, W.J., Rothman, R.H.: The noisy ceramic hip: is component malpositioning the cause? J. Arthroplasty **23**, 643–649 (2008)

32. Grimm, et al.: Annual AAOS meeting Podium No. 199 (2009)

Part III

Metal Articulations

Ceramic Surface Engineering of the Articulating Surfaces Effectively Minimizes Wear and Corrosion of Metal-on-Metal Hip Prostheses

10

Karel J. Hamelynck, David J. Woodnutt, Robin Rice, and Genio Bongaerts

10.1
Introduction

The presence of high concentrations of metal degradation particles within the hip joint and high non-physiologically serum levels of cobalt and chromium ions in the blood and organs of patients after metal-on-metal (MoM) hip arthroplasty remains a matter of concern. These concerns lead to the question whether surgeons can continue to take the risk of using MoM prostheses knowing the potential for local and systematic complications, changes and or reactions of the materials performance in MoM bearing surface articulation. It is apparent that there is sufficient reason to try and solve these problems by reducing wear particles and corrosion of the bearing surface material. The question is "how?"

Design factors have a great influence on wear of metal components, especially those factors which have an effect on lubrication [5]. Wear is minimal when fluid, like synovial fluid, separates the articulating surfaces. This situation is hard to achieve. More realistic is a combination of fluid film and mixed lubrication. In the mixed lubrication regime, the load is partially supported by a combination of contact with boundary lubricants (at the asperity tips) and by pressure developed in a fluid film that separates some, but not all, of the asperities of the articulating surfaces [5]. According to Jin et al. [6], optimal lubrication will occur and wear will be reduced when femoral head components with large diameters in combination with a smooth surface roughness and small diametrical clearances between the femoral and acetabular component are used.

K.J. Hamelynck (✉)
Van Breestraat 52, 1071ZR Amsterdam, The Netherlands
e-mail: kjhamelynck@xs4all.nl

G. Bongaerts
Orthopädisch Chirurg Gemeinschaftspraxis für OrthopädieDres, med. Jensen, Jensen, Bongaerts Rahlstedter Bahnhofstraße 7a D-22143, Hamburg, Germany

R. Rice
Consultant Orthopaedic Surgeon Clinical Director, Trauma & Orthopaedics, Nevill Hall Hospital, Brecon Road, Abergavenny Gwent NP7 7EG, Wales, United Kingdom

D.J. Woodnutt
Consultant Orthopaedic Surgeon Morriston Hospital, Heol Maes Eglwys Morriston, Swansea SA6 6NL, Wales, United Kingdom

K. Knahr (ed.), *Tribology in Total Hip Arthroplasty*,
DOI: 10.1007/978-3-642-19429-0_10, © 2011 EFORT

10

Various surgical reasons which may influence the wear of metal articulating surfaces in MoM hip arthroplasty have been identified. Femoral head components with diameters smaller than 46 mm, mostly used in women, demonstrated a revision rate much higher than femoral head components of a larger size [1]. Malposition of the acetabular component was found to be a cause of early failure [4, 7]. The common factor in these failures was lack of sufficient lubrication due to a too small a contact area or coverage angle, and edge loading of the cup, leading to excessive metal wear and subsequent negative local tissue reactions. When something was wrong within the joint, increase of metal ions in the blood was a common phenomenon [2].

Despite the fact that these surgical insufficiencies and design factors undoubtedly play a role in excessive high metal ion levels in the blood and the frequency of soft tissue reactions near the hip joint, the bearing surface itself remains the key feature of the performance of metal-on-metal components. Surprisingly the role of the material of which all metal-on-metal prostheses are made, the cobalt-chromium-molybdenum alloy (CoCrMo), is not very clear. It is now generally accepted that alloys with higher carbon content are performing better than those with low-carbon content [9]. However, no differences were found, when various conditions from as-cast to wrought, were tested [3]. Also the use of heat treatment, by which carbides are reduced in number and size, is controversial.

A new approach to try and reinforce the articulating metal surfaces is surface engineering using a ceramic. The use of ceramic surface engineering in order to reduce wear and corrosion of metals is not uncommon in non-medical application. Ceramic surface engineering is extensively used outside the human body to reduce wear of metals and to protect metals against corrosion. In the automotive industry, ceramic surface engineering is used in bearings, brakes, camshafts, cylinder heads, pistons and valve springs. Ceramic surface engineering is also used in aerospace, missile, machine tools, and constructive industry. During the surface engineering process, a ceramic is integrated into the metal surfaces by physical vapor deposition (PVD). The value of PVD technology lies in its ability to modify the surface properties of a device without changing the underlying material properties and biomechanical functionality. In addition to enhanced wear resistance, PVD coatings reduce friction, and are compatible with sterilization processes whilst increasing the materials resistance to corrosion.

A unique hard-on-hard bearing hip prosthesis system was designed for Resurfacing Hip Arthroplasty (RHA) and conventional Total Hip Arthroplasty (THA) with large femoral head components (Fig. 10.1). The articulating surfaces of the metal (CoCrMo) components of this system after the normal production process were treated with ceramic surface engineering, using the ceramic titanium-niobium-nitride (TiNbN). After an extensive period of preclinical, mechanical and biological testing this system has been used in clinical practice from 2001 in THA, and from 2004 in RHA.

10.2
Purpose of the Study

The purpose of the study was to investigate whether the increase of the chromium and cobalt ion levels, which is normally seen in the blood of patients after MoM hip arthroplasty, can be prevented by using ceramic surface engineering components. Because in conventional THA metal ions are also generated at the cone and stem of the femoral components, the study of

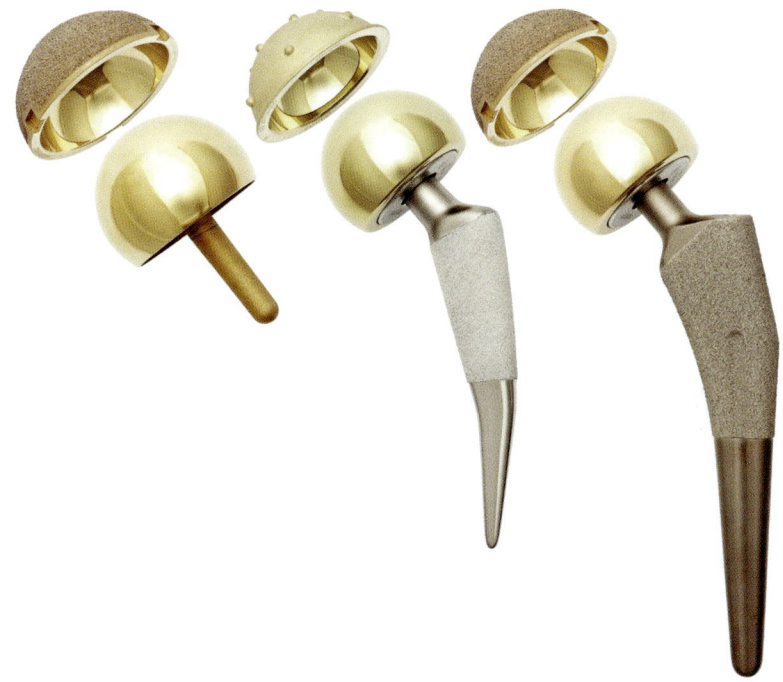

Fig. 10.1 Accis® Ceramic Surface Engineered components for total and resurfacing hip arthroplasty

metal ions in patients after RHA (no femoral stem junction) was considered to be more appropriate to demonstrate the positive effect of ceramic surface engineering.

10.3
Materials and Methods

10.3.1
Patients

Unilateral resurfacing hip arthroplasty was carried out in 200 consecutive patients from three orthopaedic centres: Morriston Hospital Swansea UK, Nevill Hall Hospital Abergavenny UK and Arthro Clinic Hamburg Germany. The surgery was carried out by one single surgeon at each centre (DW, RR, and GB respectively). One-third of the total group were women and two-third were men. The mean age at the time of the operation was 55 years (34–72 years).

10.3.2
Materials

In all patients the ACCIS® resurfacing hip hard-on-hard bearing prosthesis, was used. The metal components of the ACCIS® resurfacing hip prostheses are made from a casted chromium-cobalt-molybdenum (CoCrMo) alloy according to ISO 5832–4. After the casting phase, the components are cooled, heat treated, polished and undergo micro-surface finish. Different from the manufacture of MoM prostheses, after these treatments the articulating surfaces are then engineered with the ceramic titanium-niobium-nitride (TiNbN) by physical vapour deposition (PVD). The TiNbN is integrated into the articulating surfaces. The layer thickness is 0.3–0.9 µm. The ACCIS® prostheses for total hip arthroplasty (THA) and resurfacing hip arthroplasty (RHA) are manufactured by implantcast GmbH, Buxtehude, Germany. The ceramic surface engineering process is performed at DOT (Dünnschicht und Oberflachen Technik) in Rostock, Germany. The acetabular component is designed for press-fit cementless fixation with a triple-radius outside geometry with equatorial widening design and has a cementless backside coating of pure Titanium plasma spray (TPS) according to the ISO standard 5832–2.

10.3.3
Methods

Blood samples of 60 randomly selected patients with unilateral RHA, 20 from each centre, were taken and analyzed before surgery and at intervals of 3, 6, 12 and 24 months after surgery. Independent trace metal measurements were performed at the Universitätsklinikum Carl Gustav Carus, Dresden, Germany.

10.4
Results

The results showing "chromium and cobalt ion concentrations in the blood of patients after ACCIS® RHA up to 2 years after surgery" are shown in Fig. 10.2.

Pre- surgery the median chromium concentrations were 0.5 (0.25–2.8) and the median cobalt concentrations were 0.82 (0.075–2.86) µgr/L. The average chromium concentration was 0.589 and the average cobalt concentration was 0.903 µgr/L.

The median chromium concentrations after 2 years were 0.94 (0.25–3.6) and the median cobalt concentrations after 2 years were 1.04 (0.07–2.36) µgr/L. The average chromium concentration after 2 years was 1.073 and the average cobalt concentration after 2 years was 1.095 µgr/L.

None of the patients post operation has shown to have any apparent increase of Co and Cr ions in their blood, at 2 years' post-surgery.

10.5
Discussion

Metal ion measurement is a valuable tool for diagnosis and patient follow-up after MoM hip arthroplasty. Serum ion concentrations of cobalt and chromium can be used to estimate the amount of wear taking place in these devices [2]. Normal ion levels published in the

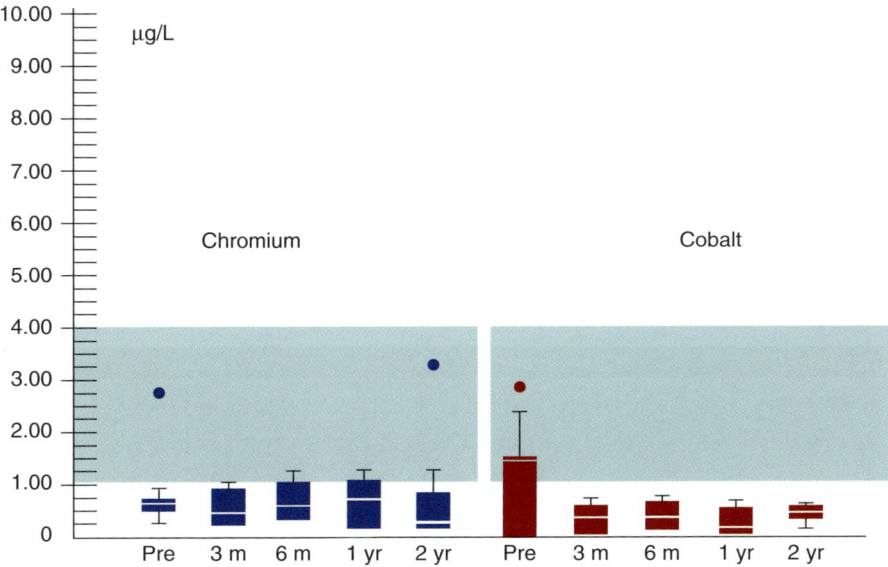

Fig. 10.2 Box plot showing metal ion data in μg/L. as published by MacDonald [8] The inferior, middle and superior horizontal lines of the boxes represent the first quartile, median and third quartile. The ends of the whiskers correspond to the limits of the data, beyond which values are considered anomalous. The mean is displayed with a-, outliers with a° and extreme outliers with an*. The light blue zone 0, 5-4 μg/L indicates the "normal" levels as described in the Handbook for Environmental Medicine

Table 10.1 Normal metal ion levels in patients without an implant [10]

	Serum	Whole blood
Chromium	<1.0 μg/L	0.5–4.0 μg/L
Cobalt	<3.9 μg/L	0.5–3.9 μg/L

Handbook of Environmental Medicine [10] are found to vary in whole blood from 0.5 to 4.0 for chromium and from 0.3 to 3,9 μgr/L : see Table 10.1.

Serum cobalt and chromium measurements are generally higher in patients after MoM hip arthroplasty than in patients without a prosthesis. However, when measuring the metal ion levels, it is important to measure the levels before surgery first, because there are great differences between pre-surgery values of patients from different regions. In this study, differences were found between pre-surgery values of the patients from Wales and Hamburg. Witzleb [11] published much lower pre-operative metal ion levels in patients from Dresden, Germany. All measurements were performed in the same laboratory in Dresden: see Table 10.2.

In this study of metal ion levels after RHA using a hard-on-hard prostheses with ceramic surface engineered bearing surfaces, the metal ion levels didn't show any change during the first 2 years of post-surgery. Most wear in MoM prostheses is usually seen during the

Table 10.2 Varying pre-surgery metal ion values in different areas

	Cobalt	Chromium
Swansea (Wales UK)	0.96 µg/L	1.87 µg/L
Abergavenny (UK)	0.81 µg/L	0.84 µg/L
Hamburg (Germany)	0.99 µg/L	1.31 µg/L
Dresden (Germany)	0.50 µg/L	0.25 µg/L

Table 10.3 Correlation between head size and mean metal ion levels after 2 years

Head size	Chromium	Cobalt
42–44 mm	1.26 µg/L (0.92–1.80)	1.08 µg/L (0.90–1.18)
46–48 mm	0.63 µg/L (0.25–0.94)	1.14 µg/L (0.62–1.59)
>50 mm	0.91 µg/L (0.25–3.68)	1.08 µg/L (0.52–2.29)

running-in period between the first 500,000 and 2,000,000 million cycles of motion [5]. During this running-in phase, surface asperities formed by block-carbides are worn flat. After this period, wear of the articulating surfaces becomes minimal and corrosion of the articulating surfaces is the more important phenomenon. The absence of any increase in the metal ion levels is also in great contrast to the published data of other MoM prostheses. Because the ion level is believed to be a diagnostic tool to identify clinical complications, the absence of any increase of the metal ion levels demonstrates that wear of the ACCIS® Ceramic Surface Engineered components has been minimal. Corrosion does not occur as a result of the protection that ceramic surface engineering offers.

The correlation between the mean metal ion levels, 2 years after the index operation, and femoral head size is summarized in Table 10.3. There was no difference in results between hips with smaller or larger femoral head size.

The influence of cup position had not been measured in all patients, when this presentation was given at the EFORT Congress 2010. However, the mean metal ions were normal in seven patients with a steep position of the cup (five patients with more than 50° and two patients with more than 55° of abduction): chromium 1.64 (0.62–2.82 and cobalt 1.77 (0.62–4.59) µgr/L. Data about the cup position and metal ion levels in the blood of a greater cohort of patients after ACCIS® RHA will be published this year.

The conclusion of this new study about the effect of cup position on blood metal ions is known already: in none of the cases with components placed in a less than optimal position an increase in Co and or Cr metal ion levels was demonstrated. Whilst malposition of the cup plays an important role in producing metallosis within the hip joint and increased metal ion levels in the blood of patients after MoM RHA without ceramic surface engineered surfaces, no change of metal ion levels was seen in patients with the ACCIS® Ceramic Surface Engineered hard-on-hard bearing hip prosthesis. The ceramic surface engineering evidently minimizes wear and provides excellent protection against corrosion.

10.6
Conclusion

– Ceramic surface engineering of metal articulating surfaces of MoM hip prostheses effectively minimizes wear and corrosion and thus metal ion release.
– The absence of any increase of metal ion levels indicates that metal wear is minimal.

References

1. Australian Orthopaedic Register Yearbook (2008)
2. De Smet, K., De Haan, R., Calistri, A., Campbell, P.A., Ebramzadeh, E., Pattyn, C., Gill, H.S.: Metal ion measurement as a diagnostic tool to identify problems with metal-on-metal hip resurfacing. J. Bone Joint Surg. Am. **90**(Suppl 4), 202–208 (2008)
3. Dowson, D., Hardaker, C., Flett, M., Isaac, G.H.: A hip simulator study of the performance of metal-on-metal joints. Part I: the role of materials. J. Arthroplasty **19**(8), 118–123 (2004)
4. Grammatopoulos, G., Pandit, H., Glyn-Jones, S., et al.: Optimal acetabular orientation for hip resurfacing. J. Bone Joint Surg. Br. **92**(8), 1072–8 (2010)
5. Isaac, G.H., Thompson, J., Williams, S., Fisher, J.: Metal-on-metal bearing surfaces: materials, manufacture, design, optimization, and alternatives. Proc. Inst Mech. Eng H J. Eng. Med. **220**(2), 119–133 (2006)
6. Jin, Z.M., Dowson, D., Fisher, J.: Analysis of fluid film lubrication in artificial hip joint replacements with surfaces of high elastic modulus. Proc. Inst. Mech. Eng. H. J. Eng. Med. **211**(3), 247–256 (1996)
7. Langton, D.J., Jameson, S.S., Joyce, T.J., Webb, J., Nargol, A.V.: The effect of component size and orientation on the concentrations of metal ions after resurfacing arthroplastyof the hip. J. Bone Joint Surg. Br. **90**(9), 1143–51 (2008)
8. MacDonald, S.J., Brodner, W., Jacobs, J.J.: A consensus paper on metal ions in metal-on-metal hip arthroplasties. J. Arthroplasty **19**(Suppl 2), 12–16 (2004)
9. Streicher, R.M., Semtitsch, M., Schon, R., Weber, H., Rieker, C.: Metal-on-metal articulation for artificial hip joints: laboratory study and clinical results. Proc. Inst. Mech. Eng. H. J. Eng. Med. **210**(3), 223–232 (1996)
10. Wichman, et al.: "Handbuch der Umweltmedizin" *ISDN : 978-3-609-71180-5* (2007)
11. Witzleb, W.C., Ziegler, J., Krummenauer, F., Neumeister, V., Guenther, K.P.: Exposure to chromium, cobalt, molybdenum from metal-on-metal total hip replacement, hip resurfacing arthroplasty. Acta Orthop. **77**(5), 697–705 (2006)

Retrieval Wear Analysis of Metal-on-Metal Hip Resurfacing Implants Revised Due to Pseudotumours

11

Young-Min Kwon, Harinderjit S. Gill, David W. Murray, and Amir Kamali

11.1
Introduction

With the introduction of the metal-on-metal (MoM) bearing to hip resurfacing, there has been a recent and rapid increase in the number of hip resurfacing procedures performed. Metal-on-metal bearings have superior wear properties than conventional metal-on-polyethylene (MoP) bearings [1, 2]. This has the potential to substantially reduce wear-induced osteolysis as the major cause of failure. Other proposed advantages of metal-on-metal hip resurfacing arthroplasty (MoMHRA) over conventional total hip arthroplasty (THA) include bone conservation, greater implant stability, and assumed easier revision surgery [3]. The data from the National Joint Registry of England and Wales and the Australian National Joint Replacement Registry have indicated that hip resurfacing procedures accounted for 46% and 29%, respectively, of all primary hip arthroplasty procedures performed in the patient group younger than 55 years of age [4, 5]. However, there is a growing concern regarding the occurrence of periprosthetic soft-tissue lesions in patients with MoMHRA, causing significant symptoms [6] and requiring revision surgery in a high proportion patients, the outcome of which is poor [7]. As it was often difficult to distinguish morphologically from a necrotic tumour, the term 'inflammatory pseudotumour' has been used to describe these solid or cystic periprosthetic soft-tissue lesions

Y.-M. Kwon (✉)
Adult Reconstructive Surgery Unit, Department of Orthopaedic Surgery, Massachusetts General Hospital, Harvard Medical School, 55 Fruit Street, GRJ 1126, Boston, MA 02116, USA
e-mail: kwon.young-min@mgh.harvard.edu

A. Kamali
Implant Development Centre (IDC), Smith and Nephew Orthopaedics Ltd, Aurora House, Spa Park, Harrison Way, Leamington Spa, CV31 3HL, UK
e-mail: amir.kamali@smith-nephew.com

H.S. Gill
Nuffield Department of Orthopaedics, Rheumatology and Musculoskeletal Sciences, University of Oxford, Headington OX3 7LD, UK

D.W. Murray
Nuffield Department of Orthopaedics, Rheumatology and Musculoskeletal Sciences, University of Oxford, Nuffield Orthopaedic Centre, Headington OX3 7LD, UK

K. Knahr (ed.), *Tribology in Total Hip Arthroplasty*,
DOI: 10.1007/978-3-642-19429-0_11, © 2011 EFORT

around MoMHRA [8]. The soft-tissue lesions have also been described by other names such as bursae [9], cysts [10] or inflammatory masses [11].

The presence of such soft-tissue pseudotumours in patients with MoMHRA has been associated with elevated serum and hip aspirate levels of cobalt (Co) and chromium (Cr), the principal elements in the CoCr alloy used in MoMHRA implants [6, 12]. Patients with pseudotumours had up to a six-fold elevation of median serum levels and up to a 12-fold elevation of median hip aspirate levels of Co and Cr ions in comparison to patients without pseudotumours [12]. Blood concentrations of blood cobalt and chromium have been considered as potential surrogate markers of metal-on-metal (MoM) bearing surface wear, as it is not possible to measure the wear of bearing couples radiographically in metal-on-metal hip systems. This assumption is supported by recent wear measurements of retrieved MoM implants [13]. Elevated serum metal ion concentrations were significantly correlated with greater amounts of femoral component linear wear. As wear is positively correlated with the elevation of metal ion concentrations in vivo, this suggests that pseudotumours in patients with contemporary MoMHRA implants are associated with increased wear at the MoM articulation. Furthermore, edge-loading, a phenomenon whereby the femoral component comes into contact with the edge of the acetabular component, has been recently suggested as a possible mechanism that leads to increased wear in MoMHRA implants [13–16]. However, direct assessment of wear and occurrence of edge-loading in MoMHRA implants has not been previously investigated in patients with pseudotumours.

The aims of this study were therefore firstly to quantify the wear of MoMHRA implants revised for pseudotumours using direct measurement and compare this measurement to that for a control group of MoMHRA implants revised for other reasons of failure, and secondly, to establish whether the edge-loading phenomenon occurs in implants revised for pseudotumours.

11.2
Materials and Methods

11.2.1
Selection of Retrieved Implants

The current study aimed to investigate two groups of retrieved implants: (1) Group 1 – MoMHRA implants retrieved from patients who underwent revision surgery due to the presence of pseudotumour; and (2) Group 2 (control) – MoMHRA implants retrieved from patients who underwent revision surgery due to other causes of failure. The study needed to be performed on representative samples in each implant group to ensure a valid generalisation of the findings. This was particularly important in selecting implants for the pseudotumour group, because the presence of pseudotumour is not a well-established revision indication. In order to minimise selection bias, the retrieved implants were selected using the following selection criteria:

1. Inclusion criteria included: (a) MoMHRA implants of a contemporary design that are currently in use. This was imposed to reflect the design of MoMHRA implants in patients with pseudotumours in whom the elevated metal ion concentrations were measured

[6, 12]; (b) MoMHRA revision surgery diagnosis of pseudotumour was confirmed by meeting the following requirements: (i) the presence of soft-tissue pseudotumour lesion was documented in the revised resurfaced hip pre-operatively; (ii) the stated indication for revision surgery was solely due to the presence of pseudotumour; and (iii) the final diagnosis of the histopathology report of the tissue specimen by the musculoskeletal pathologist was pseudotumour, in accordance with the published histological features of pseudotumour [6, 8]; and (c) MoMHRA revision surgery diagnosis in the non-pseudotumour group was limited to femoral neck fracture and infection. Femoral neck fractures and infection represented two MoMHRA failures that could be reliably established from the operative notes and/or histopathology reports. Other reasons for failure such as component loosening can have multiple underlying causes and thus, the diagnosis is often difficult to establish.

2. Exclusion criteria included: (a) MoMHRA implants that had been severely damaged at removal, as these would have led to unreliable roundness measurement and hence wear estimation; (b) MoMHRA implants in the pseudotumour group which showed discrepancy between the operative indication and the histopathology report. This exclusion ensured that diagnosis of pseudotumour was confirmed by medical records, operative notes as well as the final histopathology report in all MoMHRA implants in the pseudotumour group.

11.2.2
Collection of Retrieval Implants

Approval from the local Research Ethics Committee was obtained prior to accessing implants from the archived holdings at the authors' institution, which contained retrieved MoMHRA implants obtained from revision surgeries performed. In total, 30 MoMHRA implants were investigated. There were eight MoMHRA implants retrieved from the patients who underwent revision surgery due to the presence of pseudotumour (Group 1) and 22 MoMHRA implants retrieved from the patients who underwent revision surgery due to femoral neck fractures or infection (Group 2). A summary of the implants in each group is provided in Table 11.1. Although seven out of eight implants in the pseudotumour

Table 11.1 Summary of retrieved implants

	Pseudotumour group	Non-pseudotumour group
Number of implants	8	22
Gender (female:male)	8:0	13:9
Age (years)	52 (range 39–65)	54 (range 45–70)
Mean time in vivo (years)	3.6 (range 1.1–6.6)	2.3 (range 1.0–5.8)
Implant type	BHR (5) Conserve plus (2) Cormet (1)	BHR (18) Conserve plus (4)
Mean femoral component size (mm)	47 (range 42–50)	48 (range 44–54)

group contained both the acetabular and the femoral components, the acetabular components were not available for analysis in four femoral neck fracture cases as they were not always replaced when the femoral component was revised.

11.2.3
Study Power Estimation

Due to a limited number of available retrieved implants that met the selection criteria, it was important to estimate the power of the study. This was done using the Altman's nomogram [17]. For linear wear rate of MoMHRA, a significant difference of interest was 10 μm, which represented a two-fold increase from the reported steady-state wear rate of 5 μm for MoM bearings [2]. Based on repeated measures taken during a pilot study on five implants, the expected measurement standard deviation was estimated to be 8 μm. Thus, a standardised difference of 1.25 was calculated. The statistical significance level was set at 0.05. Using Altman's nomogram [17], the sample size in the current study ($n=30$) estimated the power at 0.90. This implied that the study had 90% probability of correctly concluding that a difference of 10 μm exists. However, as the nomogram is used under the assumption that equal sized samples are required, the unequal sized samples in this study would have reduced the power.

11.2.4
Linear Wear Assessment Using a Roundness Machine

All implants were catalogued with an identification number to ensure patient anonymity. The components were not autoclaved between removal and examination. The bearing surfaces of the retrieved components were first inspected with the naked eye under bright lighting. The linear wear of retrieved MoMHRA femoral and acetabular bearing surfaces was assessed using a Taylor-Hobson Talyrond 290 Roundness machine (Taylor Hobson Ltd., Leicester, United Kingdom) in a blinded fashion at the Implant Development Centre (IDC), Leamington Spa, UK.

   The roundness testing machine measures both partial and full circles. The machine was used to measure the roundness of the femoral and acetabular components in several planes. Multiple equatorial roundness profiles were then taken from the skirt of the femoral component and the edge of the acetabular component toward their polar regions to locate the maximum wear on each component. This was done in a sequential manner in 5 mm increments. The wear area was identified as a discontinuity in the characteristic manufactured profile. Once the wear area was located, further equatorial roundness profiles were taken in 1 mm increments away from the area in order to identify the maximum wear value. Polar measurements perpendicular to the equatorial axis were also taken at the maximum wear area. The location of the wear scar (maximum depth of wear) was recorded in degrees from the centre of polar region on the prosthesis. The maximum linear wear, recorded in micrometres (μm), was measured by subtracting the wear profile from an ideal circle. For each MoMHRA femoral and acetabular component, the average linear wear rate was defined as the maximum linear wear scar depth divided by the duration of the implant in vivo (years).

11.2.5
Edge-Loading Definition

Edge-loaded components were defined as acetabular cups which showed the maximum area of wear crossing over the edge of the cup (Fig. 11.1). Non-edge-loaded components were defined as acetabular cups which showed the maximum wear area on the cup occurring within the hemispheric bearing surface of the cup.

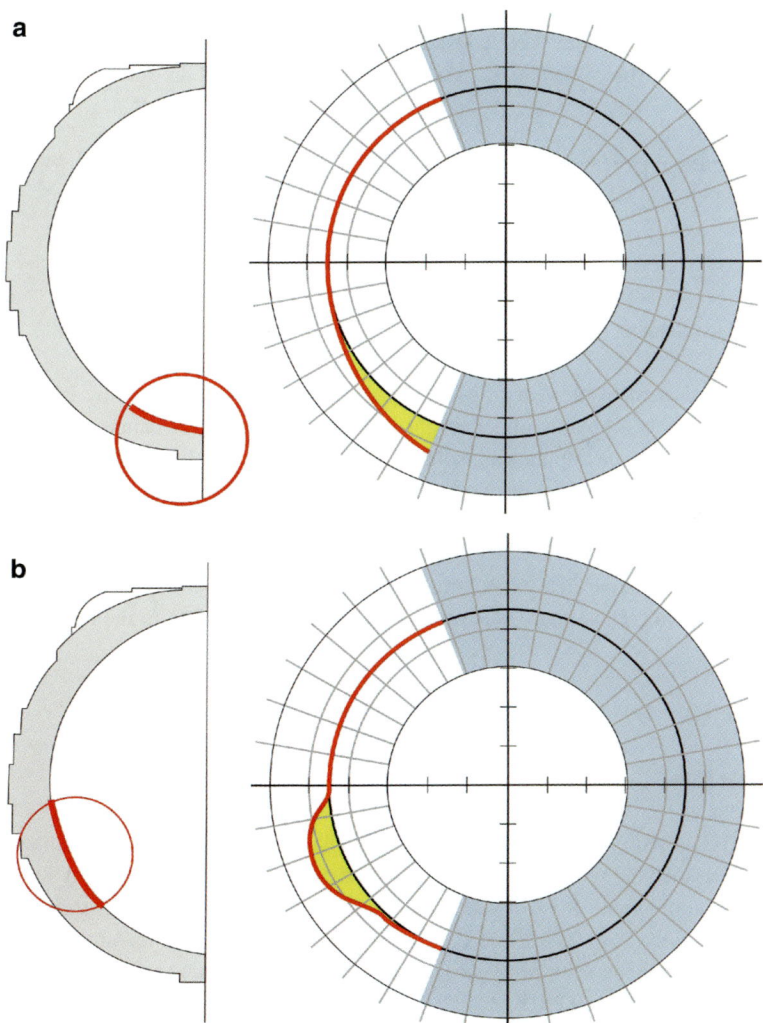

Fig. 11.1 Roundness profiles of (**a**) edge-loaded and (**b**) non-edge loaded acetabular cup components. The maximum wear area (*shaded in yellow*) is located at the edge of the cup for the edge-loaded (**a**), while the maximum wear area occurs well within the cup in the non-edge loaded cups (**b**)

11.2.6
Statistics

The data set was assessed for normality. Mann–Whitney non-parametric tests were used to calculate the level of statistical significance for the differences in the non-normally distributed linear wear and wear rate between the pseudotumour and non-pseudotumour implant groups. Pearson's correlation analyses, expressed as Pearson's correlation coefficient, were performed to assess the strength of correlation between the linear wear rate and time to revision. The incidence of edge-loading in each group was compared using the Fisher's exact test. Differences at $p<0.05$ were considered to be significant. SPSS® statistical software release 13.0 (SPSS Inc. Chicago, IL, USA) was used to perform the statistical analyses.

11.3
Results

11.3.1
Qualitative Assessment

Qualitative assessment of wear of MoMHRA implants was performed by inspecting the bearing surfaces of implants under bright lighting. Post-retrieval damage artefacts were evident in a number of implants as isolated scratches or dents. In all cases, fine scratches were observed on the polar regions of bearing surfaces of the femoral head and the acetabular cup. There was no evidence of stripe wear such as that reported in ceramic-on-ceramic bearing couples [18]. This may be explained by the ability of MoM bearings to self-polish by wearing down the surface surrounding a scratch, which may hide any distinctive visible stripe on the head. Thus, there was minimal wear damage visible to the naked eye in all implants.

11.3.2
Linear Wear Rate

In comparison with the non-pseudotumour implant group, the pseudotumour implant group was associated with (Fig. 11.2): (1) significantly higher median linear wear rate of the femoral component: 8.1 μm/year (range 2.75–25.4 μm/year) vs. 1.79 μm/year (range 0.82–4.15 μm/year), $p=0.002$; and (2) significantly higher median linear wear rate of the acetabular component: 7.36 μm/year (range 1.61–24.9 μm/year) vs. 1.28 μm/year (range 0.18–3.33 μm/year), $p=0.001$.

Similarly, differences were also measured in absolute wear values. The median absolute linear wear was significantly higher in the pseudotumour implant group: (1) 21.05 μm (range 2.74–164.80 μm) vs. 4.44 μm (range 1.50–8.80 μm) for the femoral component, $p=0.005$; and (2) 14.87 μm (range 1.93–161.68 μm) vs. 2.51 μm (range 0.23–6.04 μm) for the acetabular component, $p=0.008$. In all cases, the maximum wear occurred in localised zones.

Fig. 11.2 Boxplots showing differences in the linear wear rates of the (**a**) femoral head and (**b**) acetabular cup components between the two implant groups

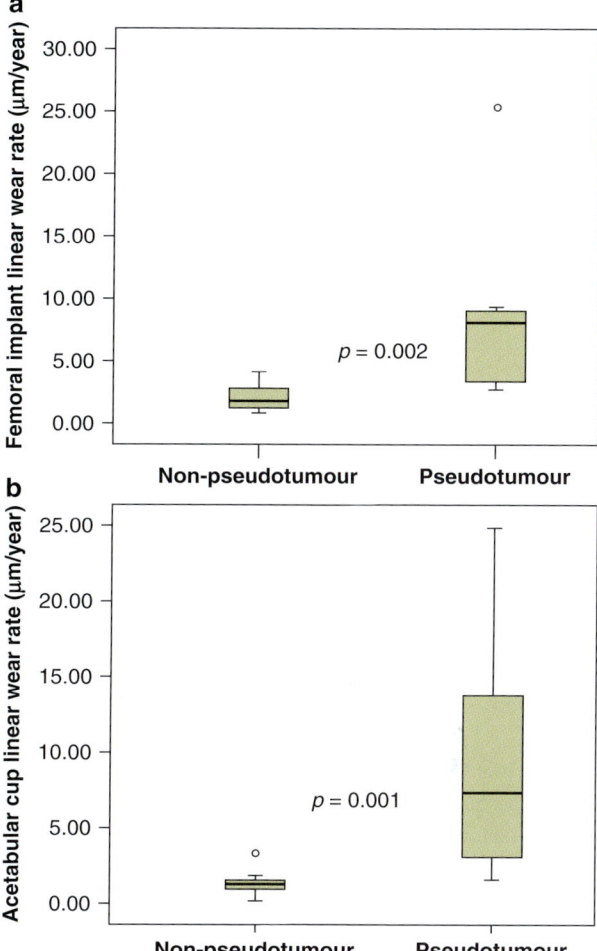

There was a poor linear correlation between the linear wear rate and the time in vivo: Pearson's correlation coefficient (r): $r=0.33$ for femoral component, and $r=0.32$ for acetabular component (Fig. 11.3). In addition, there was no significant difference in linear wear rate between the fracture and infection sub-groups within the non-pseudotumour group ($p=0.41$).

11.3.3
Edge-Loading

Using the definition of edge-loading previously described (Fig. 11.1), edge-loading was observed in all acetabular components (100%) in the pseudotumour group. In contrast, edge-loading was observed in only one acetabular component (6%) in the non-pseudotumour

Fig. 11.3 Linear wear rates of all retrieved implants are plotted against time in vivo

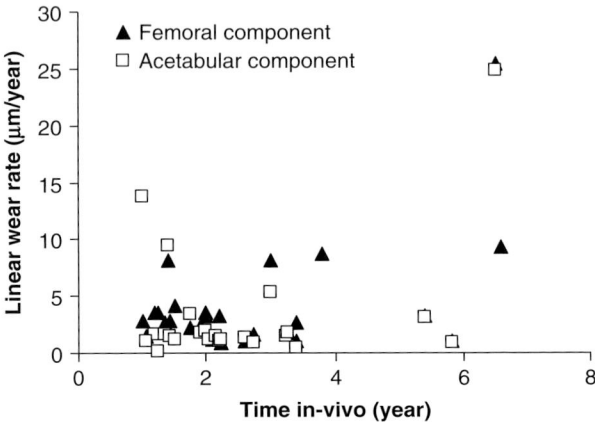

group of implants. The deepest wear was observed well within the bearing surface for the rest of the non-pseudotumour group. The difference in the incidence of edge-loading between the two groups was statistically significant (Fisher's exact test, $p=0.03$).

11.4
Discussion

Wear debris from MoM bearing surfaces is generated by mechanical wear, surface corrosion or a combination of both. It consists of both insoluble particles and metal ions, the latter disseminating into the systemic circulation [19]. MoMHRA patients with pseudotumours were found to have up to a six-fold elevation of median serum levels and up to a 12-fold elevation of median hip aspirate levels of cobalt and chromium ion levels in comparison to patients without pseudotumours [6, 12]. In fact, pseudotumours were not detected in those patients who had normal cobalt and chromium levels. In light of the recently reported positive correlation between the elevated serum metal ion concentrations and the greater amounts of MoMHRA implant wear [13], these study results suggested that pseudotumours occur when there is increased wear at the MoM articulation. However, direct wear measurement of MoMHRA implants has not been previously investigated in patients with pseudotumours.

The results of the current retrieval study demonstrated that MoMHRA implants revised due to pseudotumour were associated with significantly greater linear wear of both femoral head and acetabular cup components compared to a control group of implants revised for other reasons of failure. Implants in the pseudotumour group had up to a four-fold increase of median linear wear rate of the femoral component and greater than a five-fold increase of median linear rate of the acetabular component, in comparison with implants in the non-pseudotumour group.

Interestingly, the median linear wear rates of the femoral components measured in the pseudotumour group in this study were, however, similar to the wear rates recently reported by Witzleb et al. in eight BHR femoral components revised for non-pseudotumour-related failures [20]. Seven out of eight implants in the report by Witzleb et al. were revised within 15 months of surgery. It is difficult to compare these values as the values determined from short-term retrievals during the run-in phase are higher than the steady state values. In fact, a five-fold reduction in annual wear rate from the first year run-in phase to the third year steady-state has been measured in MoM bearings [2]. The mean time in vivo for both implant groups in the current study was during or beyond the third year of implantation (3.6 years for the pseudotumour group and 2.3 years for the non-pseudotumour group). Furthermore, a variation in the methods of measurement used between the current study (roundness machine) and the short term published report (coordinate measurement machine) may have also contributed to differing estimations of the absolute linear wear rates. However, these factors do not diminish the differences found in relative values between the two implant groups in this study.

Potential confounding factors that may have led to increased wear rates in the pseudo-tumour implant group include relative differences in time from implantation to retrieval, femoral component size and heterogenicity of subgroups in the non-pseudotumour group. However, there was a poor correlation between the linear wear rate and the time in vivo for implants in both groups. Furthermore, there was no significant differences in femoral component sizes between the pseudotumour and non-pseudotumour groups (47 mm vs. 49 mm, $p=0.75$). Lastly, the linear wear rates between the two sub-groups (femoral neck fracture and infection) in the non-pseudotumour implants did not differ significantly ($p=0.41$). Therefore, significantly greater linear wear rates measured from the retrieved MoMHRA implants revised due to pseudotumour support the in vivo study findings of elevated metal ion concentrations in serum and hip aspirates in patients with pseudotumours [6, 12], thereby confirming that pseudotumour in contemporary MoMHRA implants is associated with increased wear at the MoM articulation.

The morphology of the wear patch on the acetabular cup components in the pseudo-tumour group was always characterised by maximum wear being at the edge of the cup surface, indicating edge-loading of the bearing. This suggested that the contact zone of the bearing was predominantly positioned at the edge of the cup during the component life. Thus, it is likely that the wear mechanism was not one of a normal wear patch expanding with time over the edge of the cup, but rather one in which the head component was inadequately covered by the cup component from the moment of initial weight bearing. Edge-loading has been recently suggested as a possible mechanism that leads to increased wear in MoMHRA implants [13, 14, 16, 21] because it has the potential to disrupt the favourable fluid film lubrication that occurs in MoMHRA bearing surfaces [22, 23]. As edge-loading and the thickness of the lubricating fluid-film influence the extent of metal ion generation [14], these conditions would lead to increased metal ion levels in the hip joint aspirate and serum, such as those measured in patients with pseudotumours. Therefore, edge-loading with an associated loss of fluid film lubrication may be the dominant wear generation mechanism in MoMHRA patients with pseudotumour.

There are several limitations to the study. Linear wear measurement method is limited to estimating the local deepest wear (worst case) and not the mean wear of the component. However, wear was found to be highly localised in the MoMHRA implant couples.

The topographical surface shape of the wear scar was not mapped. It would have been informative to fully characterise the corresponding wear scars located on the femoral components. Anatomical orientation of the implants is difficult without landmarks provided at the time of revision surgery as MoMHRA implants, unlike stemmed THA, are symmetrical in appearance without distinctive geometrical markers that can be used for orientation. As the retrieved implants were not marked with in vivo orientation at the time of retrieval, information regarding the specific anatomical orientation of the wear scar could not be assessed. Despite these limitations, the pattern of edge-loading was clearly demonstrated on the acetabular components. Complete good-quality plain radiographs were only available in fewer than half of the selected retrieved implants. Therefore, radiographic acetabular component inclination, which has been reported to influence MoM bearing wear [21, 24], could not reliably be assessed.

It was assumed in the current study that the rate of wear is linear with time. Data from hip simulator studies and clinical metal ion level measurement suggest a biphasic rate of wear with an initial high run-in wear followed by a lower steady state wear [25, 26]. Although the vast majority of implants in both groups were from or beyond the third year of implantation, several implants in the non-pseudotumour group had failed earlier within the higher wearing run-in period. This may have led to under-estimation of the wear difference between the implant groups. The wear rates are also sensitive to material parameters (carbon content, alloy processing and heat treatment), design parameters (radial clearance), and manufacturing parameters (surface roughness and sphericity) [25, 27]. There are differences in designs, materials, and manufacturing processes between three types of metal-on-metal hip resurfacing implants measured in the current study. This confounding factor could be controlled if only one type of implant from a single manufacturer was measured. However, this would have further reduced the available number of retrieved implants, thereby reducing the power of study. Moreover, a proportionally similar number of implant types were studied in each group, and thus minimising this confounding effect on comparative values.

In conclusion, significantly higher linear wear rates were measured in the retrieved femoral head and acetabular cup components revised due to pseudotumour than those revised for other reasons in the current study; this supports the in vivo study findings of elevated metal ion concentrations in patients with pseudotumours [6, 12]. The results of this study, therefore, provides the first direct evidence to confirm that pseudotumour in contemporary MoMHRA implants is associated with increased wear at the MoM articulation. Thus, soft-tissue pseudotumour, a clinical complication with a high revision burden, may represent a local biological reaction to increased wear debris burden, generated by excessive MoMHRA implant wear. Furthermore, highly localised wear occurring due to edge-loading with an associated loss of fluid film lubrication may be the dominant wear mechanism in MoMHRA patients with pseudotumours as no pseudotumours were observed in non edge loaded devices. Further in vivo investigations are required to evaluate the risk factors of edge-loading such as component positioning as a mechanism responsible for increased wear in MoMHRA patients. Such evidence-based knowledge would be pivotal for patient selection, surgical technique and design of future hip resurfacing implants in order to ensure long-term prosthesis survivorship.

References

1. MacDonald, S.J., McCalden, R.W., Chess, D.G., Bourne, R.B., Rorabeck, C.H., Cleland, D., Leung, F.: Metal-on-metal versus polyethylene in hip arthroplasty: a randomized clinical trial. Clin. Orthop. Relat. Res. **406**, 282–296 (2003)
2. Sieber, H.P., Rieker, C.B., Kottig, P.: Analysis of 118 second-generation metal-on-metal retrieved hip implants. J. Bone Joint Surg. Br. **81**(1), 46–50 (1999)
3. Shimmin, A., Beaule, P.E., Campbell, P.: Metal-on-metal hip resurfacing arthroplasty. J. Bone Joint Surg. Am. **90**(3), 637–654 (2008)
4. National Joint Registry England and Wales Annual Report (2008)
5. Australian Orthopaedic Association National Joint Replacement Registry Annual Report (2008)
6. Pandit, H., Glyn-Jones, S., McLardy-Smith, P., Gundle, R., Whitwell, D., Gibbons, C.L., Ostlere, S., Athanasou, N., Gill, H.S., Murray, D.W.: Pseudotumours associated with metal-on-metal hip resurfacings. J. Bone Joint Surg. Br. **90**(7), 847–851 (2008)
7. Grammatopoulos, G., Pandit, H., Kwon, Y.M., Gundle, R., McLardy-Smith, P., Beard, D.J., Murray, D.W., Gill, H.S.: Hip resurfacings revised for inflammatory pseudotumours have a poor outcome. J. Bone Joint Surg. Br. **91**(8), 1019–1024 (2009)
8. Pandit, H., Vlychou, M., Whitwell, D., Crook, D., Luqmani, R., Ostlere, S., Murray, D.W., Athanasou, N.A.: Necrotic granulomatous pseudotumours in bilateral resurfacing hip arthoplasties: evidence for a type IV immune response. Virchows Arch. **453**(5), 529–534 (2008)
9. Campbell, P., Shimmin, A., Walter, L., Solomon, M.: Metal sensitivity as a cause of groin pain in metal-on-metal hip resurfacing. J. Arthroplasty **23**(7), 1080–1085 (2008)
10. Gruber, F.W., Bock, A., Trattnig, S., Lintner, F., Ritschl, P.: Cystic lesion of the groin due to metallosis: a rare long-term complication of metal-on-metal total hip arthroplasty. J. Arthroplasty **22**(6), 923–927 (2007)
11. Boardman, D.R., Middleton, F.R., Kavanagh, T.G.: A benign psoas mass following metal-on-metal resurfacing of the hip. J. Bone Joint Surg. Br. **88**(3), 402–404 (2006)
12. Kwon, Y.M., Ostlere, S., McLardy-Smith, P., Gundle, R., Whitwell, D., Gibbons, C.L.M., Athanasou, N.A., Gill, H.S., Murray, D.W.: Metal ion levels in pseudotumours associated with metal-on-metal hip resurfacings. In: The 55th Orthopaedic Research Society Annual Meeting, Las Vegas, USA, (2009)
13. De Smet, K., De Haan, R., Calistri, A., Campbell, P.A., Ebramzadeh, E., Pattyn, C., Gill, H.S.: Metal ion measurement as a diagnostic tool to identify problems with metal-on-metal hip resurfacing. J. Bone Joint Surg. Am. **90**(Suppl 4), 202–208 (2008)
14. Campbell, P., Beaule, P.E., Ebramzadeh, E., LeDuff, M., De Smet, K., Lu, Z., Amstutz, H.C.: The John Charnley Award: a study of implant failure in metal-on-metal surface arthroplasties. Clin. Orthop. Relat. Res. **453**, 35–46 (2006)
15. De Haan, R., Pattyn, C., Gill, H.S., Murray, D.W., Campbell, P.A., De Smet, K.: Correlation between inclination of the acetabular component and metal ion levels in metal-on-metal hip resurfacing replacement. J. Bone Joint Surg. Br. **90**(10), 1291–1297 (2008)
16. Hussain, A., Counsell, L., Kamali, A.: Clinical effects of edge loading on metal-on-metal hip resurfacings. British Hip Society, Manchester, United Kingdom (2009)
17. Altman, D.G.: Statistics in Practice. British Medical Association, London (1982)
18. Taylor, S., Manley, M.T., Sutton, K.: The role of stripe wear in causing acoustic emissions from alumina ceramic-on-ceramic bearings. J. Arthroplasty **22**(7 Suppl 3), 47–51 (2007)
19. Jacobs, J.J., Gilbert, J.L., Urban, R.M.: Corrosion of metal orthopaedic implants. J. Bone Joint Surg. Am. **80**(2), 268–282 (1998)
20. Witzleb, W.-C., Guenther, K., Hanisch, U., Ziegler, J., Guenther, K.-P.: In vivo wear rate of the Birmingham Hip Resurfacing arthroplasty. J. Arthroplasty **24**(6), 951–956 (2009)

21. De Haan, R., Pattyn, C., Gill, H.S., Murray, D.W., Campbell, P.A., De Smet, K.: Correlation between inclination of the acetabular component and metal ion levels in metal-on-metal hip resurfacing replacement. J. Bone Joint Surg. Br. **90**(10), 1291–1297 (2008)
22. Udofia, I.J., Jin, Z.M.: Elastohydrodynamic lubrication analysis of metal-on-metal hip-resurfacing prostheses. J. Biomech. **36**(4), 537–544 (2003)
23. Liu, F., Jin, Z., Roberts, P., Grigoris, P.: Importance of head diameter, clearance, and cup wall thickness in elastohydrodynamic lubrication analysis of metal-on-metal hip resurfacing prostheses. Proc. Inst. Mech. Eng., H: J. Eng. Med. **220**(6), 695–704 (2006)
24. Langton, D.J., Jameson, S.S., Joyce, T.J., Webb, J., Nargol, A.V.: The effect of component size and orientation on the concentrations of metal ions after resurfacing arthroplasty of the hip. J. Bone Joint Surg. Br. **90**(9), 1143–1151 (2008)
25. Dowson, D., Hardaker, C., Flett, M., Isaac, G.H.: A hip joint simulator study of the performance of metal-on-metal joints: part I: the role of materials. J. Arthroplasty **19**(Suppl 8), 118–123 (2004)
26. Back, D.L., Young, D.A., Shimmin, A.J.: How do serum cobalt and chromium levels change after metal-on-metal hip resurfacing? Clin. Orthop. Relat. Res. **438**, 177–181 (2005)
27. Chan, F.W., Bobyn, J.D., Medley, J.B., Krygier, J.J., Tanzer, M.: The Otto Aufranc Award. Wear and lubrication of metal-on-metal hip implants. Clin. Orthop. Relat. Res. **369**, 10–24 (1999)

Part IV

Polyethylene Articulations

Polyethylene Wear in Total Hip Arthroplasty for Suboptimal Acetabular Cup Positions and for Different Polyethylene Types: Experimental Evaluation of Wear Simulation by Finite Element Analysis Using Clinical Radiostereometric Measurements

12

Christian Wong and Maiken Stilling

Abbreviations

2D	two-dimensional
3D	three-dimensional
CoCr	cobalt-chromium
DSS	directional shear stress
EGS models	elementary geometric shape models
E-poly	E-vitamin stabilized
FE	finite element
HCL	highly cross-linked
PE	polyethylene
RMSE	root mean square error
RSA	radiostereometric analysis
SD	standard deviation
SI	stress intensity
THA	total hip arthroplasty
UHMWPE	ultra high molecular weight polyethylene
VMS	Von Mises stress

C. Wong (✉)
Department of Orthopaedics, Hvidovre Hospital, Kettegårds Allé 30, 2650 Hvidovre, Denmark
e-mail: chwo123@gmail.com

M. Stilling
Orthopaedic Research Unit, Aarhus University Hospital, Tage-Hansens Gade 2, 8000 Aarhus C, Denmark
e-mail: maiken.stilling@ki.au.dk

K. Knahr (ed.), *Tribology in Total Hip Arthroplasty*,
DOI: 10.1007/978-3-642-19429-0_12, © 2011 EFORT

12.1
Introduction to Polyethylene Wear and Wear Measurement Methods

12.1.1
Polyethylene Wear: Is It Important?

Innovations in polyethylene (PE) processing and improved PE biomechanics over the past decade have sparked clinical as well as industrial interest in early and independent product documentation. PE wear particles initiate biological responses that contribute to peri-prosthetic osteolysis and in longer terms result in aseptic loosening and prosthetic revision surgery [11, 16, 30], and therefore high-wear bearings could potentially induce an increased secondary revision burden. Radiological methods for early estimation of clinical PE wear are often used [27], but at least some years of follow-up is required to predict the future consequences, and in general the results of clinical controlled series are not awaited prior to a broader commercial introduction. This is despite the fact that past innovations of total hip arthroplasty include many examples of unexpected – but late – recognition of implant failures related to the enthusiastic embracement of new implant products prior to proper independent evaluation as it was the case with hylamer liners [10, 13, 14, 22]. Methods using experimental computer simulations of PE wear and stress parameters in the liner could be used as a valuable preclinical and independent evaluation of the isolated mechanical improvements in new PE liners.

12.1.2
Experimental PE Wear Simulation and Finite Element Analysis

Computational analysis methods such as finite element (FE) analysis are useful in theoretical prediction of material performances, as an alternative, where wear simulation studies clinical trials are lacking or not yet performed. FE analysis is a numerical technique for finding approximate solutions of partial differential equations as well as of integral equations using computers, and FE allows detailed visualization of where structures bend or twist, and indicates the distribution of stresses and displacements [2]. The FE method provides a "biomechanical estimate" for several structural parameters such as fracture, fatigue, and creep. The method can also be use to examine mechanical parameters correlated to volumetric wear in relation to, i.e., implant position, body weight, and material types. Ideally, FE "wear simulation" should be the initial approach for estimating volumetric wear of new bearing materials before proceeding with further experimental and clinical evaluation.

Earlier studies using the FE analysis method generated mechanical parameters based on a function of roughness [8], sliding velocity [28], or sliding distance [24, 25], which have been regarded as proportional contributors to volumetric wear. These wear parameters were validated in comparison to in vitro experimental wear studies [7, 17, 19, 24, 25]. Furthermore, contact pressure and cross-shear motion is associated with polyethylene wear [6, 8] and have been incorporated in studies of FE wear analysis [7, 26], and recently, functions of contact pressure and cross shear have been correlated to wear by Kang et al. [18, 19].

It is important that computational methods are evaluated against by experimental data or clinical data [43, 44]. For FE wear analysis, this has been done by comparing FE wear results to multi-directional pin-on-plate testers [17, 19] and in-hip simulators [24]. However, the discrepancy between wear predictions and experimental data was large and up to 4.1% [25] or a factor 2–3 in difference [3, 19]. This disagreement raises concern for the usefulness of the predictions of FE wear results, especially because the in vitro validation studies of FE wear predictions are only valid for the applied and often a single set of loading conditions. Although in vitro testing provide important control over the testing situation and hence an accurate FE setup [28], it is obvious that an in vivo evaluation of FE wear simulation would be of interest because it include the numerous facets of actual wear. To the knowledge of the authors, a clinical validation of FE wear simulations has never been performed before. When comparing in vivo wear data to similar computed FE wear results, another approach for simulation of the in vivo loading has to be applied, since an accumulated sum of highly various loads over time have to be simulated. This has been attempted by the estimation of a "daily loading cycle" for optimization of bone structures [21, 41], however; in the experience of the authors, such assumptions are still far from reality. Instead we propose a summarized and simplified load, which appear adequate for long-term effect studies, since realistic accumulative loading over time cannot be simulated [9, 43, 44].

12.1.3
Clinical PE Wear Measurement and Radiostereometric Analysis

Where the computed optimal conditions for cup position or material improvements can be used to predict clinical PE wear over time, clinical measurements comprise the advantage that it includes all wear-provoking factors, i.e., patient differences, surgery, materials, clinical use, sterilization method, cup position, cup design, coating of the metal backing, creep, impingement, back-side wear, PE locking mechanism, liner thickness, femoral head-size, third-body wear, gender, age, activity level, height, and weight.

Radiological PE wear measurement is essentially a measurement of the femoral head penetration into the metal cup using different mathematical equations and different types of radiographs. All methods have in common that they cannot distinguish between creep (slow material deformation but without particle production), bedding-in (the backside of the PE liner wearing into a higher conformity with the metal shell), running-in (the initial fitting of the femoral head into the polyethylene liner), back-side wear, and true wear [38]. The pattern of PE wear is typically high in the first period after surgery and then decreases with time. The reason is that the femoral head penetrates into the acetabular polyethylene due to a combination of creep and wear, and that bedding-in and running-in results in larger contact surfaces with lower contact stresses and lower rates of wear [29, 37]. Creep decreases over time and is considered to be important within the first 6–12 months, after which PE wear is described as linear (the true wear) or in a "steady state" [38].

The clinical measure of interest in PE wear analysis is the femoral head penetration over time (wear rate) and, in some cases, the determination of steady-state wear rates. Several methods are available and useable but radiostereometric analysis (RSA) is the most accurate and the gold standard, but also the most costly method [32, 33]. The methodology of RSA is using two synchronized roentgen tubes to obtain two simultaneous radiographs of a patient positioned over a calibration box. From these images, and using three-dimensional (3D) surface models in relation to other models or bone markers, the motion of the center of gravity (femoral head) can be calculated according to a fixed rigid body reference (metal cup) [32, 33].

In vivo PE wear is multidirectional, and clinical methods that assume a single direction of wear typically underestimate the true amount of wear [45–47]. Accurate steady-state femoral head penetration rates can be used to estimate when complete liner wear-through will likely happen, and this may be useful to determine how frequently a patient should return for follow-up examinations – or when to schedule the patient for revision surgery. However, specific phenomena may change the head penetration patterns and obscure the linearity of radiographic wear, i.e., third-body wear debris and changes in the surface smoothness of bearing surfaces and these factors are not encountered in FE Analysis.

12.2
Study A: FE Wear Simulation

12.2.1
Finite Element (FE) Analysis Investigating Cup Position and Liner Quality

Cup positioning in total hip arthroplasty is believed to be crucial for the risk of long-term wear hence aseptic loosening. In this sub-study, we conducted FE analysis of an acetabular PE liner to examine the mechanical consequences related to polyethylene wear with optimal and sub-optimal positions of the acetabular cup.

Secondly, we carried out FE analysis of an UHMWPE, HCL, and E-poly liner to examine the mechanical behavior related to polyethylene wear, and subsequently the mechanical performance of these liner types in suboptimal positions.

12.2.2
FE Wear Simulation

A finite element model of a PE liner of the Ringloc type (Biomet Inc.) was analyzed in this study. We used similar wear parameters as in earlier studies [7, 17, 19, 24, 25], but modified them to contact surface VMS, SI, and DSS, since a singular load was used in a comparative study for the different cup positions and material types of PE.

12.2.2.1
FE Mesh

FE analyses were performed using the finite element code in Solidworks™. The FE model was a 3D model of a PE liner from an acetabular cup. The geometric t-dimensional model for the polyethylene liner to fit the Ringloc acetabular metal backing was provided by courtesy of Biomet. An FE mesh of the liner was generated with a total of 10,960 4-node volumetric tetrahedral elements. Convergence test for the FE model was then performed to ensure, that the FE model had an appropriate number of elements. When this test is performed, the number of elements in the FE model is increased until a certain point, where the calculated results converge to the one exact solution, thus giving the appropriate number of elements to use for the FE model. See Fig. 12.1 for the FE mesh of the liner.

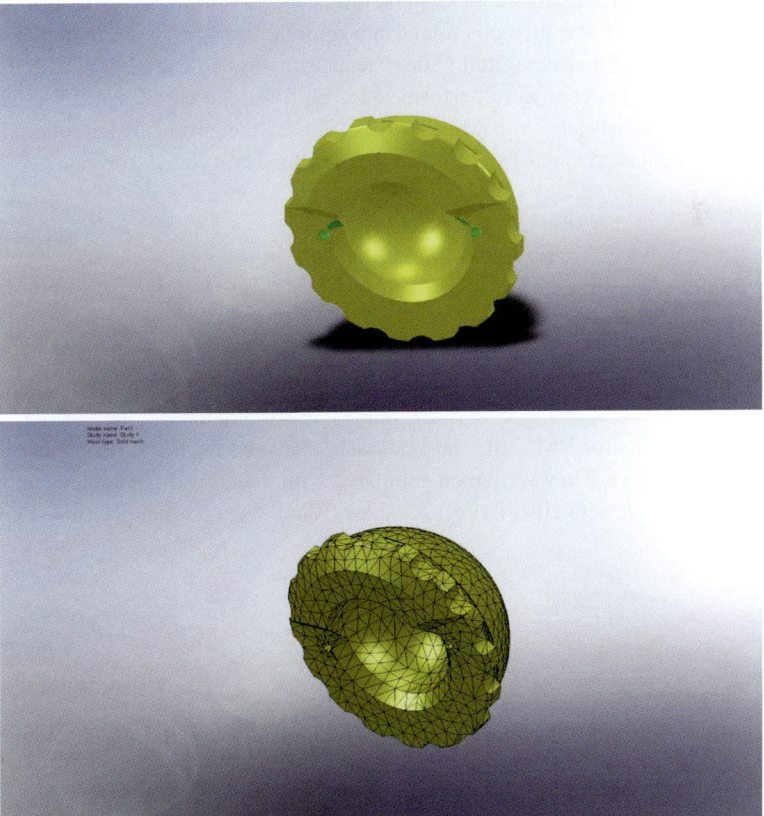

Fig. 12.1 Graphical and FE mesh model of an acetabular liner

12

12.2.2.2
Material Properties

The material properties of Young's modulus and Poisson's ratio for three types of PE liner were analyzed. These data were provided with courtesy of Biomet [1], but will not disclosed in this book. The material properties of UHMWPE (Arcom), HCL (Arcom XL), and E-vitamin stabilized (E-Poly) polyethylene were applied to the FE mesh of the polyethylene liner.

12.2.2.3
Loading and Boundary Conditions

Four types of loads were applied to the FE model of the PE liner. The loads were identical to contact forces of the femoral head of the prosthesis. Assumptions of no micromotion in the metal backing and initial fully osseo-integration of the cup was made by completely immobilizing the backside of the PE liner in all directions. The first load was applied to the PE liner positioned in a standard position in 15° anteversion and 45° inclination. The load of gravity for an 80 kg person was applied to the FE model in a standing position. The PE liner was then positioned sub-optimally with 5° to coronal plane (ante/retroverted), with 10° to the sagittal plane (steep) and aligned completely with the axial plane (flat). Static analyses were performed for all four load cases. The FE model was evaluated by nodal surface Von Mises stress, Stress Intensity, and Directional Shear Stress. These structural stress parameters were assumed correlated to wear.

12.2.3
FE Wear Results

12.2.3.1
Static Analysis

The static FE results for VMS, SI, and DSS were analyzed for the various cup positions and for material types. They were then compared with the normal positioned cup and the UHMWPE liner, respectively. Figure 12.2 shows VMS for a normal positioned cup with UHMWPE and HCL PE.

12.2.3.2
Cup Positions

The maximal VMS for a steep cup was three times higher (for all the tested materials) and the minimal VMS was 60% times higher when compared to a normal positioned cup. Figure 12.3 shows VMS for a steep positioned cup.

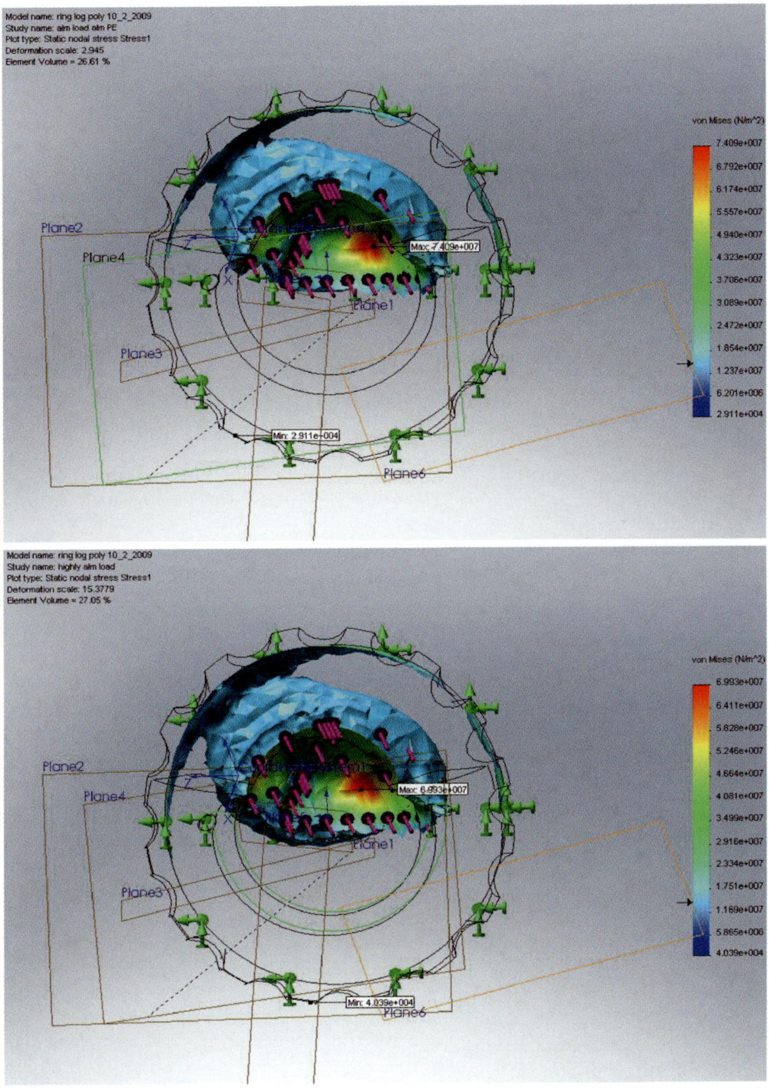

Fig. 12.2 Von Mises stress for a normal positioned cup of UHMWPE and HCL PE

For the other suboptimal positions, there were no marked differences in VMS. Figure 12.4 shows VMS for a flat positioned cup.

Table 12.1 shows the ratio of maximal VMS for the various cup positions compared to a normal cup.

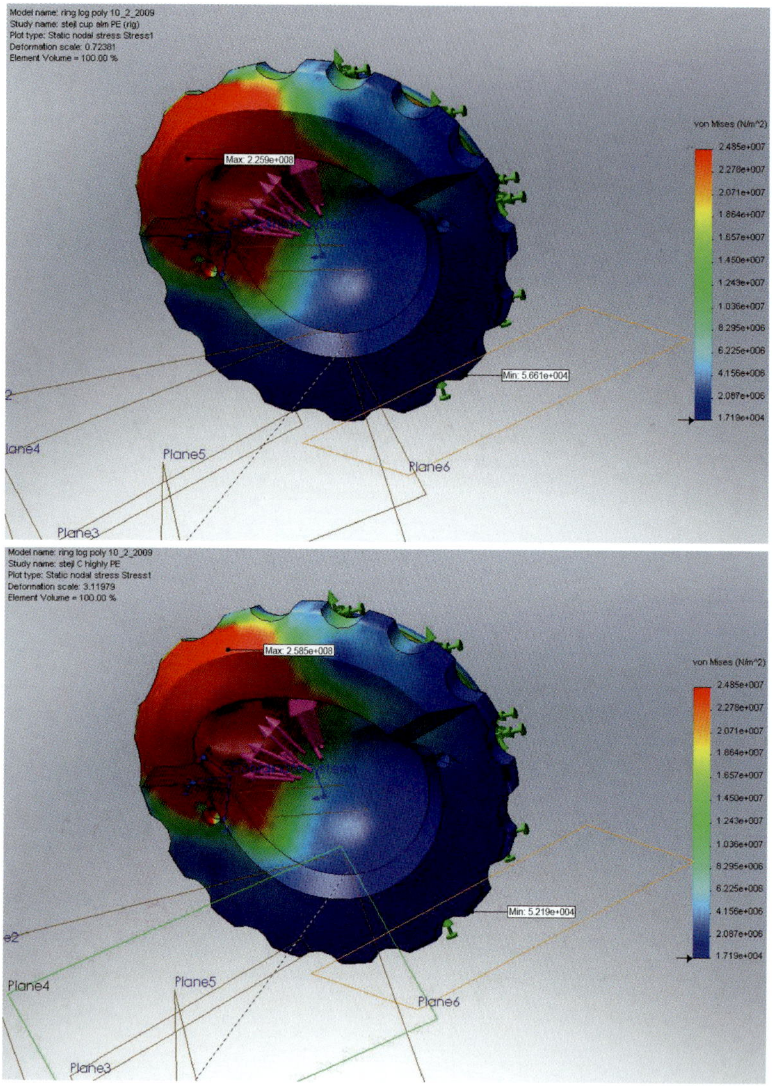

Fig. 12.3 Von Mises stress for a steep positioned cup of UHMWPE and HCL PE

12.2.3.3
Material Differences

The results for VMS, SI, and DSS in the PE liner were compared for the tested material types. For the steep cup VMS was 3.4% better for the HCL and 7% better for the E-poly PE, when compared to the UHMWPE. Figure 12.5 shows VMS for a normal positioned cup with UHMWPE and E-poly.

Figure 12.6 shows VMS for a steep cup with UHMWPE and E-poly.

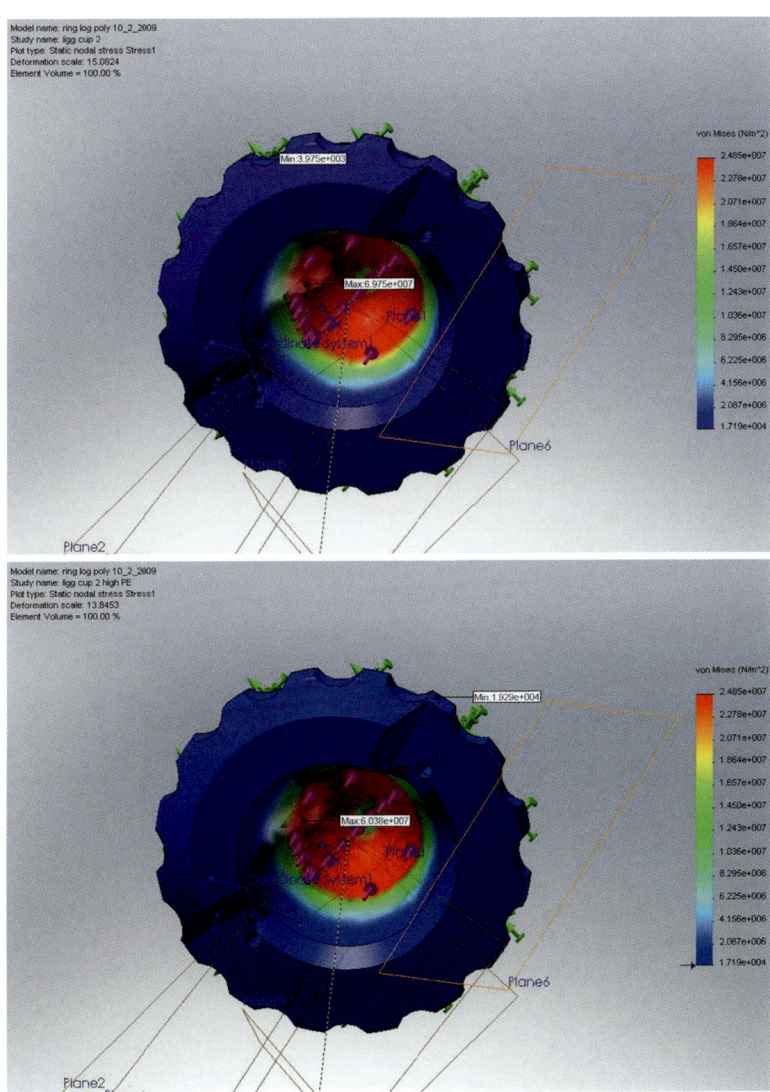

Fig. 12.4 Von Mises stress for a flat positioned cup of UHMWPE and HCL PE

Table 12.1 Ratio of maximal Von Mises stress for the various cup positions compared to 45° cup position

Cup position	UHMWPE	HCL PE	E poly
Steep	3.323	3.233	3.090
Flat	0.959	0.955	0.955
Ante/retroverted	0.858	0.858	0.858

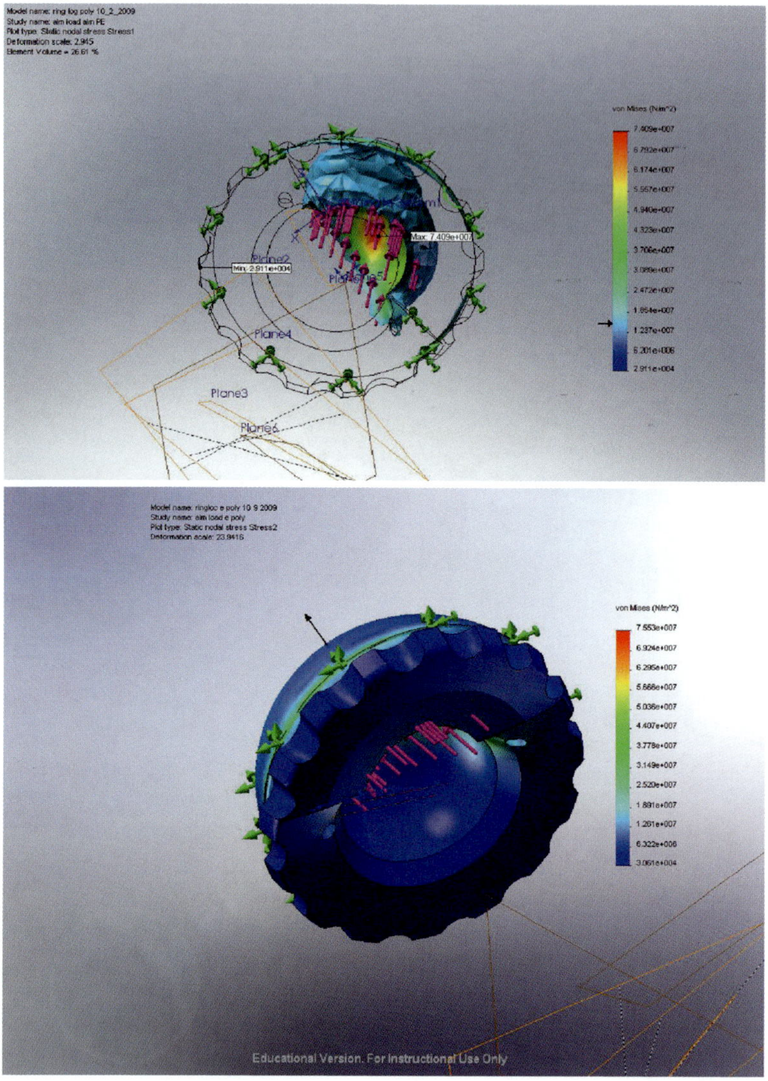

12

Fig. 12.5 Von Mises stress for a normal positioned cup of UHMWPE and E-poly PE

Figure 12.7 shows VMS for flat positioned and ante/retroverted cup with E-poly.
There were no marked differences in SI and DSS for all material types and cup positions.

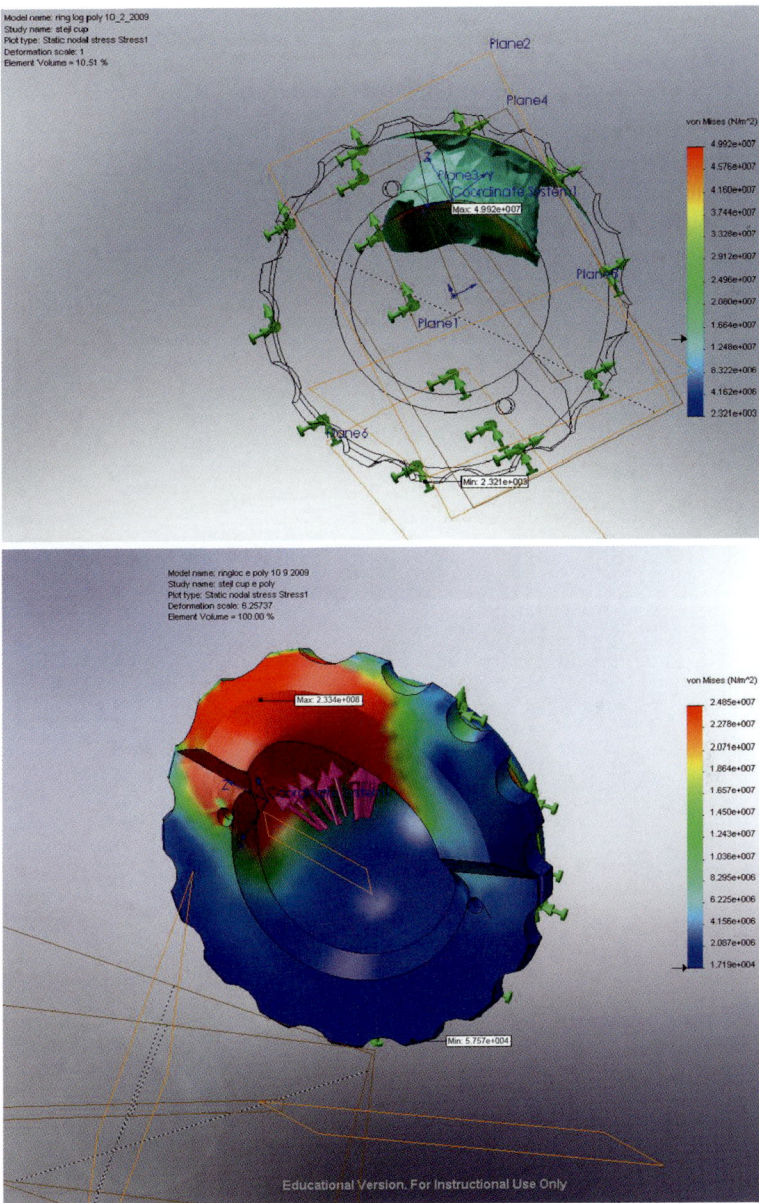

Fig. 12.6 Von Mises stress for a steep positioned cup of UHMWPE and E-poly PE

12

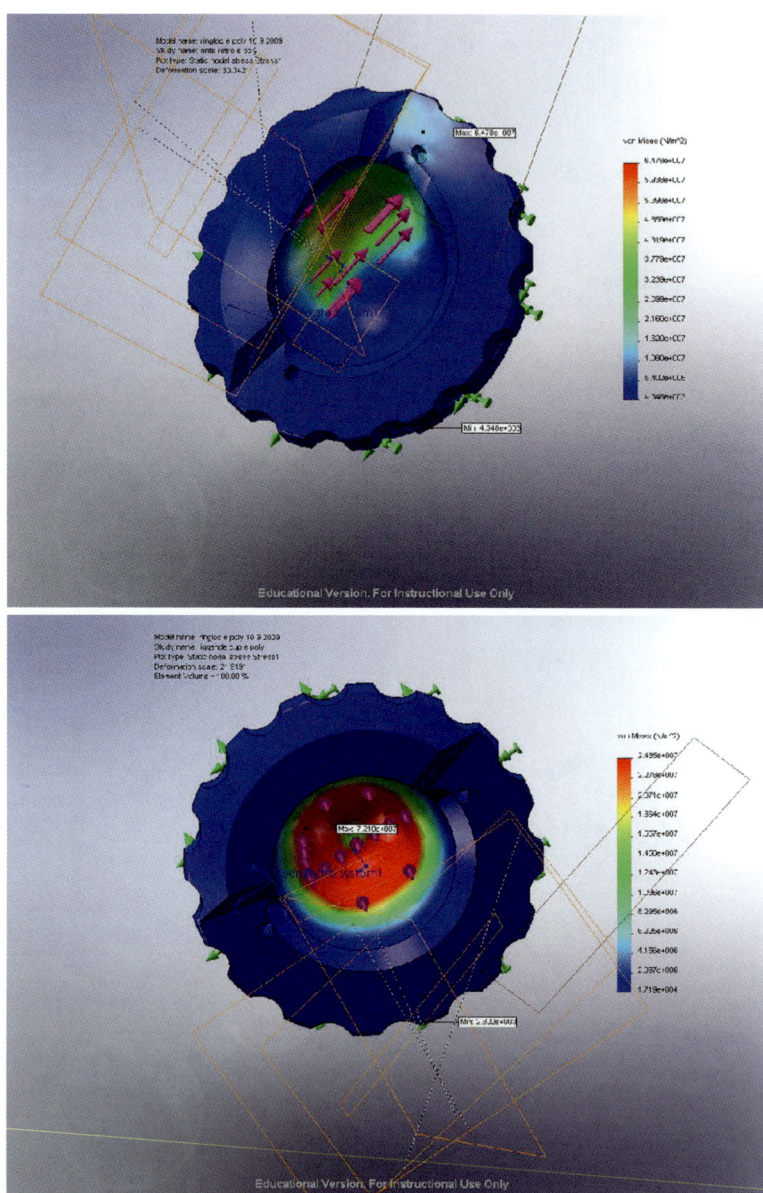

Fig. 12.7 Von Mises stress for a flat and ante/retroverted positioned cup of E-poly PE

Table 12.2 shows the ratio between an E-poly and HCL PE liner compared to a UHMWPE liner for maximal VMS for the various cup positions.

Table 12.3 shows the ratio between the UHMWPE and the HCL PE liner for maximal and minimal VMS for the various Cup positions.

Table 12.2 Ratio between HCL PE and E-poly vs. UHMWPE liner for maximal VMS in the various cup positions

Cup position	HCL PE	E poly PE
Steep	0.957	0.93
Flat	1	0.955
Ante/retro-verted	1	1

Table 12.3 Ratio between UHMWPE and HCL PE for maximal and minimal VMS for the various cup positions

Cup position	Maximal VMS	Minimal VMS
Normal	1.059	0.721
Steep	0.874	1.075
Flat	0.896	1.075
Ante/retro-verted	1.152	0.206

12.2.4
FE Wear Discussion and Conclusion

It is hereby confirmed that a sub-optimally positioned cup is a key factor for wear. A steep positioned cup gives localized stress-rise at stance, hence localized wear, which in turn increases the risk of aseptic loosening and revision [16]. The improved materials HCL PE and E-poly did not show improvement for SI and DSS, but marginal stress protection evaluated as VMS. However, the results of this study were based on computer generated simulations with inherent limitations that must be accounted for. A simple load in stance was applied to simulate the wear and tear of daily living. The femoral head, stem, and cup were simulated with approximated assumptions in the boundary conditions. Also, perfect in-growth and interface conditions were assumed. The biological processes of the living tissue and interaction between tissue and implant including the effects of lubrication were omitted. The time-dependant effect of head damage and degradation of the PE, which is known to increase wear clinically, were omitted. Overall, the computational predictions assumed that for example surface VMS was linear proportional to volumetric wear. This would be true for metal-on-metal surfaces, but for a metal head on a PE surface-bearing one would also have to consider the effects of plastic deformation (creep) and strain softening. Properly a purely linear relationship between aforementioned parameters would no longer be applicable. Impingement wear was inherently omitted and backside-wear was excluded from the FE wear analysis. Despite the aforementioned limitations, the purpose of this study was to compare cup positions and material types, and therefore the predictions were still considered valid since this was a comparative study, where cup positions and material types were "tested" under the same computational conditions. For example, the contact area as a wear factor could be omitted, since these were similar for all the different cup

12

positions. Therefore, focus was on the structural parameters for the specific PE materials and cup positions.

Comparison of the attained FE results to in vivo or in vitro data could be used for validation of FE wear predictions. Such in vitro experimental comparisons have evaluated various biomechanical parameters [7, 8, 17, 19, 24–26, 28] accepting large differences between computer predictions and experimental test results [19]. A validation of the structural parameters, which we used in this study, will be validated to in vivo wear measurements in the third part of this chapter.

Functional demands in the increasingly younger and more active patients feature a preference of large diameter heads. The commercially advertised increased durability of HCL PE and E-poly has until now justified the use of larger head-sizes even though PE wear is known to increase with larger head-size due to an increase in the sliding distance [12]. In this simulation study, we found the use of HCL PE and E-poly to be only marginally beneficial in relation to wear. This raises concerns of an uncritical use of larger heads in combination with, and justified by, newer PE materials. Especially when a well-tested workhorse – the cemented THA using small metal heads and conventional PE – is durable and has good long-term survival [23].

In conclusion, HCL PE and E-poly provided marginally better VMS protection than conventional UHMWPE, but similar results for key parameters as SI or DSS for all PE types. This indicates that there would be no benefits of the newer PE in a cost-benefit analysis!

12.3
Study B: RSA Wear Measurement

12.3.1
Radiostereometric Analysis (RSA)

Out of several available methods [4, 15, 20, 39, 40, 42], RSA is currently the most accurate clinical method (the gold standard) for estimation of the magnitude of relative component displacements using radiographs. Formerly, pre-study bead-marking of the polyethylene, the cup, or the acetabular bone was required to confine rigid-body segments for assessment of component migration. Using model-based RSA methods bead-marking is now obsolete for PE wear studies because computer models or optically scanned implant models can be mathematically fit to the contours of the components and afterwards defined as migrating or rigid objects (models). Figure 12.8 illustrates the model-based RSA method.

Stereo-radiographs, that is two radiographs obtained simultaneously with the tubes positioned in converging angles over the patient and in relation to a uni-planar or bi-planar calibration box, allow for the reconstruction of a 3D coordinate system and for the calculation of relative model-displacement in sequential stereo-radiographs [5]. The output is three translations (mm) along the orthogonal axes (x = medial-lateral,

Fig. 12.8 Model-based RSA
with computer model fitting

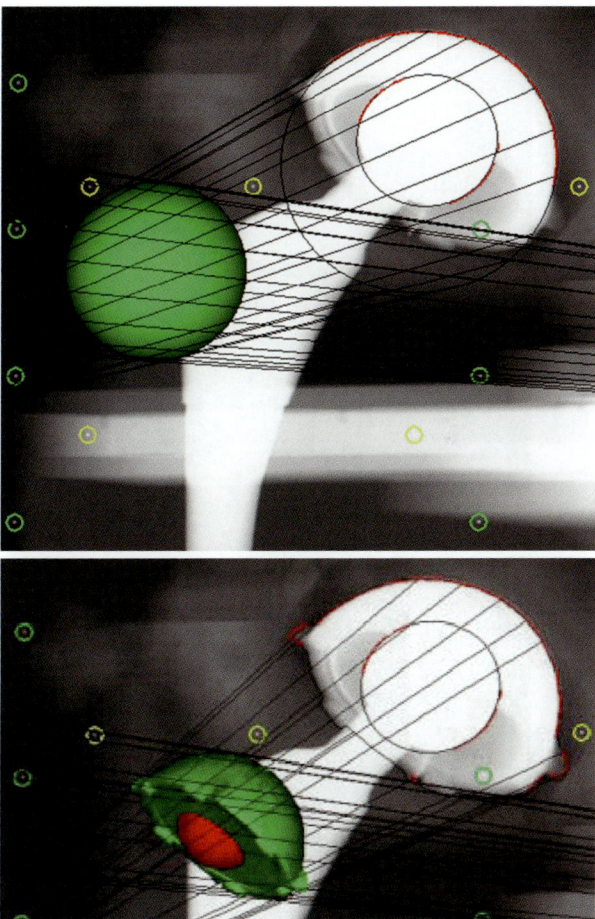

$y =$ inferior-superior, $z =$ anterior/posterior). Two-dimensional (2D) and 3D wear vectors can be calculated using Pythagoras theorem (2D wear vector = the square root of $x^2 + y^2$; 3D wear vector = the square root of $x^2 + y^2 + z^2$). The experimental set-up for RSA is shown in Fig. 12.9.

The Elementary Geometrical Shape (EGS) Model-based RSA method has formerly been shown to be highly accurate (Root Mean Square Error (RMSE) of 0.06 mm for 2D and 0.10 mm for 3D wear measurements) and precise (95% agreement limits for double measurements of 0.08 mm for 2D and 0.18 mm for 3D wear measurements) in an experimental set-up with controlled femoral head penetration in a hip phantom [36]. Accuracy (RMSE) was better for the single x and y direction (0.05 mm) compared with the z direction (0.12 mm). The experimental set-up for testing femoral head penetration is shown in Fig. 12.10.

Radiostereometric analysis using elementary geometrical shape models is shown in Fig. 12.11.

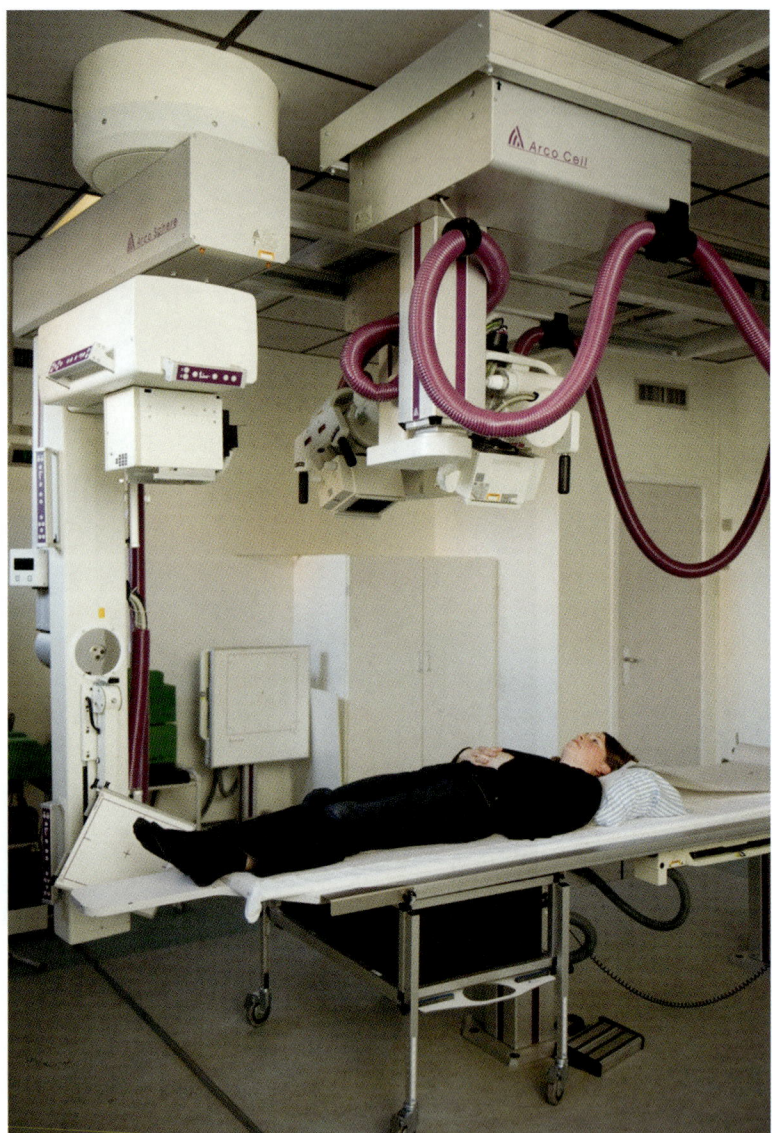

Fig. 12.9 The clinical set-up for stereo-radiography

12.3.2
Patients Included for RSA

The EGS RSA method was used clinically to describe the directional wear along the three orthogonal axes and to quantify the mean wear along the axes and the mean 2D and 3D PE wear. Liner components were in all cases ($n = 14$) a gamma-sterilized UHMWPE (ArCom,

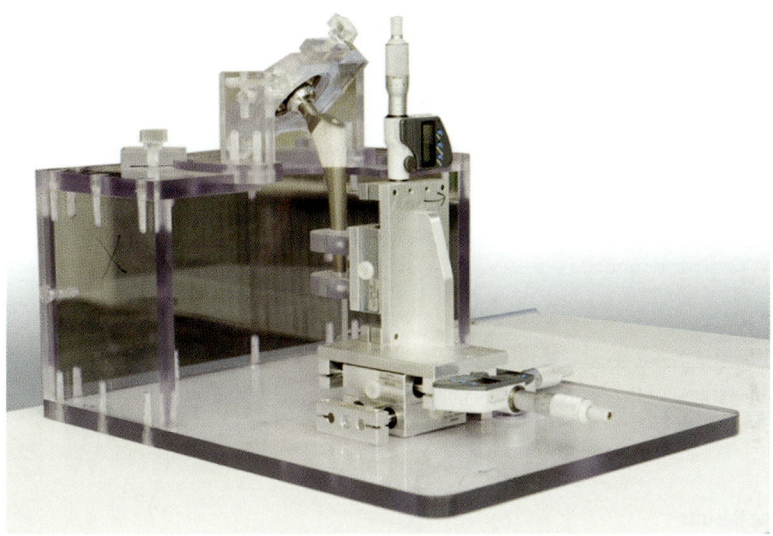

Fig. 12.10 The experimental set-up with controlled femoral head penetration in a hip phantom

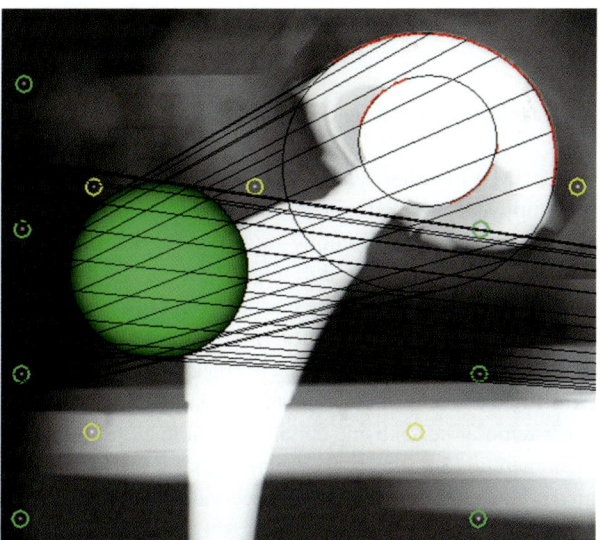

Fig. 12.11 Radiostereometric
analysis using elementary
geometrical shape models

Biomet Inc, Warsaw, IN) articulating with 28 mm Cobalt-Chromium (CoCr) femoral heads
at midterm (mean 6.1 years; range 5.3–7.1) in a young patient group (mean age 53 years;
range 44–65). The acetabular components were hydroxyapatite-coated metal shells
(Mallory-Head, Biomet Inc, Warsaw, IN). The post-operative radiographs and the final
(approximately 5 year) radiographs for each patient were used for analysis.

Table 12.4 Clinical PE wear measurements for ArCom poly (Biomet) at midterm in young patients ($n = 14$)

Analysis (wear measure)	Femoral head penetration total mean (mm)	Standard deviation (mm)	Range (mm)
X-direction wear vector	−0.09	0.27	−0.61; 0.62
Y-direction wear vector	0.53	0.30	0.05; 1.35
Z-direction wear vector	−0.19	0.51	−1.65; 0.12
2D wear ($\sqrt{x^2 + y^2}$)	0.61	0.27	0.12; 1.38
3D wear ($\sqrt{x^2 + y^2 + z^2}$)	0.77	0.38	0.21; 2.15

Directions (signs) of the coordinate axes indicate the motion of the femoral head into the acetabular cup for a right hip

12.3.3
RSA Wear Results

The experimental pooled RSA wear-data of double-examination stereo-radiographs and double-analysis in each stereo-radiograph (four analyses per patient for an averaged measure) are presented in Table 12.4.

12.3.4
RSA Wear Discussion and Conclusion

The UHMWPE liner with a standard size metal resulted in a clinical 3D wear-rate of 0.13 mm/year. The estimated wear (total mean PE wear and the PE wear rate) include creep, bedding-in, and running-in; that is PE deformity that is essentially not particle producing. A clinical threshold of interest for the detection of PE wear that leads to long-term osteolysis and implant failure has been established at 0.2 mm/year [11, 31] and the clinical wear-rate in this study is therefore supposedly "home-safe" for development particle induced osteolysis and component loosening. We have formerly evaluated the medium-term PE wear rate (0.25 mm/year) in similar UHMWPE liners articulated with 28-mm CoCr femoral heads using serial conventional radiographs and a different wear measurement method [34], where we found higher wear, but were able to show, that this relates to method bias and analysis of multiple radiographs [33, 35, 36].

The direction of wear as indicated by the signed translations was on average slightly lateral, mostly superior, and a little posterior. As specified in the ranges (see Table 12.4), the directions of wear was not uniform for all patients. We did not analyze stereo-radiographs in the period between the post-operative control and the final 5 year follow-up and therefore the "steady-state wear" could not be calculated. An approximation of the steady-state wear-rate is speculated to be about 0.1 mm/year for the investigated UHMWPE.

In conclusion, the investigated UHMWPE in a highly active patient-group does produce some wear at midterm, but the wear rate is below the critical threshold for osteolysis and long-term survival of these THAs is supposedly good as the liners were far from "worn-through."

12.4
Study C: Evaluation of FE Wear Simulations

The purpose of the third part of this tribology study was to identify a reliable FE-based structural "wear" parameter by comparison to clinical RSA wear data. To the knowledge of the authors this has never been performed previously. Although in vitro testing provide important control over the testing situation and hence an accurate FE setup [28], it is obvious that a comparison of FE wear against in vivo and accurately measured PE wear provide the most solid basis for evaluation where the numerous facets of actual PE wear are included in both analyses methods, thus making further preclinical testing using FE more valid.

12.4.1
Extended FE Wear Model

For comparison of RSA and FE wear results, a modified finite element model of a poly-ethylene liner of the Ringloc type was used. The liner position, load case, and boundary conditions for the cup were the same as earlier described, but differed in the simulation of the bearing coupling as a CoCr femoral head and neck was added to the FE model. In the literature, when in vivo PE wear data has been compared to similar computed FE results, simulation of the in vivo loading has been attempted as an estimation of a "daily loading cycle" [21, 41]. However, it is the experience of the authors that such assumptions are far from reality. Since realistic loading cannot be simulated, instead a summarized and sim-plified load, which we found adequate for long-term effect studies, was used [43, 44]. The load was still the load of gravity of an 80 kg person in a standing position, but the load was applied to the neck of the femoral stem in the FE model. The head and liner contacted each other with surface to node gap elements. Figure 12.12 shows the FE mesh of the bearing coupling.

Fig. 12.12 The FE mesh model of the bearing coupling of a THA

12

12.4.2
Evaluation of FE Wear Simulations Against Clinical RSA Wear Measurements

The FE wear results were compared to the RSA estimated resultant wear vectors (3D wear with 1 SD) projected onto the inner surface of the FE model of the liner by the three translational vectors of the orthogonal coordinate system. The circular area of 1 SD of the resultant 3D wear was then superimposed onto the liner surface. The central projected point and superimposed circular area on the liner surface were regarded as the point in an area around the liner with maximal wear. Then, the computed FE data from the inner liner surface were analyzed, and the positions of maximal contact pressure, SI, VMS, strain energy density, and total strain were identified. The distances from maximal "RSA wear" to the maximum of the above described five structural parameters were then measured.

The difference between the location of maximal wear measured by RSA and maximal contact pressure, SI, VMS, strain energy density, and total strain on the inner liner surface were all well within the range of 1 SD of the RSA data. The strain parameters of total strain and strain energy density were closest to the location of maximal RSA wear, hence they correlated the best. Figure 12.13 shows the contact pressure, SI, and VMS.

12.4.3
Evaluation of FE Wear Simulations Discussion and Conclusion

In the last part of this tribology chapter it was shown that all structural parameters for a given load were within 1 SD of the location of the maximal RSA evaluated wear, but strain parameters of *total strain* and *strain energy density* were better correlated. The comparison between simulated FE wear and in vivo RSA wear data has never been performed before, and for future studies we would add further methodological improvements. In the first part of this chapter, the FE model is discussed and improvements are suggested. Additionally, the structural, static analysis does not include factors of creep, bedding-in, running-in, and impingement wear, but focuses on the long-term important steady-state wear. Furthermore, it is realized that the direction of the single loading condition make FE wear results sensitive to load direction. For the RSA data patient variation was reflected in the large ranges for the translational wear, especially for the Z axis, which was out of plane of the cup. This raised concern for the validity of FE wear evaluation. For these reasons we did not perform an FE based estimate for yearly wear rate, since it seems that more RSA data need to be retrieved to extrapolate to the normal population, especially, since the data were collected from a young patient group. However, for future studies it would be attractive to predict the approximate time for femoral head penetration to estimate when complete liner wear-through will likely occur. This estimate would be based on yearly steady-state wear-rate obtained from FE wear analysis. This would especially be attractive if patient specific factors such as cup and stem positions could be included in such a study. This will be possible if a patient series with bead-marking of the acetabulum is available.

In conclusion, when evaluating FE based wear predictions against in vivo RSA data, the structural strain parameters seem to be the best measure for prediction of long-term steady-state wear for UHMWPE. In the first part of this chapter the stress parameters were

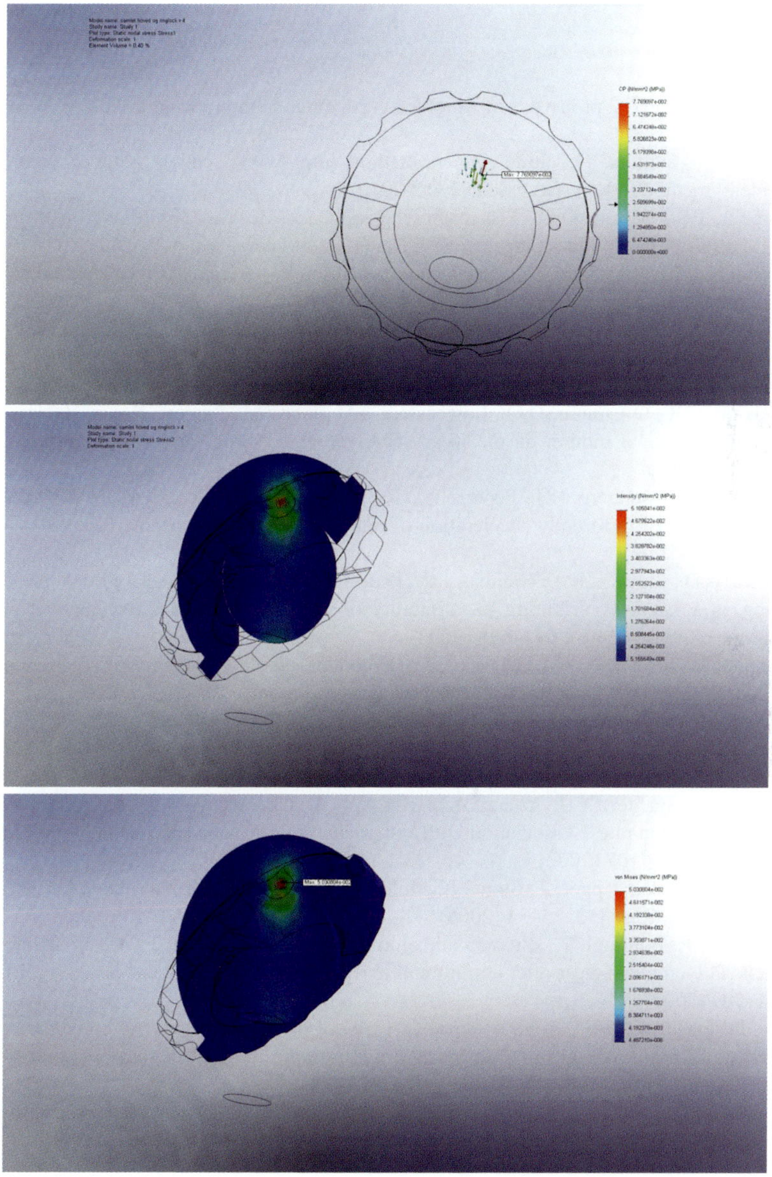

Fig. 12.13 Contact pressure, stress intensity, and Von Mises stress

used to compare wear for different types of PE, where we found no difference in predicted wear. However, since structural strain parameters later seemed to be a more reliable predictor of polyethylene wear, further FE studies are needed to confirm the hypothesis "similar wear properties of the new polyethylene materials" by including a comparison of structural strain parameters also.

12

References

1. Pappas, C.: Material properties of Ringlock cups, email communication (2009). Accessed 24 Apr 2009
2. Anonymous authors: The finite element method. http://www://en.wikipedia.org/wiki/Finite_element_method (2010). Accessed 31 Oct 2010
3. Barbour, P.S., Barton, D.C., Fisher, J.: The influence of stress conditions on the wear of UHMWPE for total joint replacements. J. Mater. Sci. Mater. Med. **8**(10), 603–611 (1997)
4. Borlin, N., Thien, T., Karrholm, J.: The precision of radiostereometric measurements. Manual vs. digital measurements. J. Biomech. **35**(1), 69–79 (2002)
5. Bragdon, C.R., Estok, D.M., Malchau, H., Karrholm, J., Yuan, X., Bourne, R., Veldhoven, J., Harris, W.H.: Comparison of two digital radiostereometric analysis methods in the determination of femoral head penetration in a total hip replacement phantom. J. Orthop. Res. **22**(3), 659–664 (2004)
6. Bragdon, C.R., O'Connor, D.O., Lowenstein, J.D., Jasty, M., Syniuta, W.D.: The importance of multidirectional motion on the wear of polyethylene. Proc. Inst. Mech. Eng. H **210**(3), 157–165 (1996)
7. Brown, T.D., Lundberg, H.J., Pedersen, D.R., Callaghan, J.J.: Nicolas Andry Award: clinical biomechanics of third body acceleration of total hip wear. Clin. Orthop. Relat. Res. **467**(7), 1885–1897 (2009)
8. Brown, T.D., Stewart, K.J., Nieman, J.C., Pedersen, D.R., Callaghan, J.J.: Local head roughening as a factor contributing to variability of total hip wear: a finite element analysis. J. Biomech. Eng. **124**(6), 691–698 (2002)
9. Cattaneo, P.M., Dalstra, M., Melsen, B.: Orthodontic tooth movement of single and multi-rooted teeth: strains and stress in the periodontium assessed by FE-analyses. **1**(1), 7. 2nd Annual Meeting of the Danish Biomechanical Society, 30 Oct 2010
10. Cohen, J.: Catastrophic failure of the Elite Plus total hip replacement, with a Hylamer acetabulum and Zirconia ceramic femoral head. J. Bone Joint Surg. Br. **86**(1), 148 (2004)
11. Dowd, J.E., Sychterz, C.J., Young, A.M., Engh, C.A.: Characterization of long-term femoral-head-penetration rates. Association with and prediction of osteolysis. J. Bone Joint Surg. Am. **82-A**(8), 1102–1107 (2000)
12. Fisher, J., Jin, Z., Tipper, J., Stone, M., Ingham, E.: Tribology of alternative bearings. Clin. Orthop. Relat. Res. **453**, 25–34 (2006)
13. Hernigou, P., Bahrami, T.: Zirconia and alumina ceramics in comparison with stainless-steel heads. Polyethylene wear after a minimum ten-year follow-up. J. Bone Joint Surg. Br. **85**(4), 504–509 (2003)
14. Huddleston, J.I., Harris, A.H., Atienza, C.A., Woolson, S.T.: Hylamer vs conventional polyethylene in primary total hip arthroplasty: a long-term case-control study of wear rates and osteolysis. J. Arthroplasty **25**(2), 203–207 (2009)
15. Hurschler, C., Seehaus, F., Emmerich, J., Kaptein, B.L., Windhagen, H.: Comparison of the model-based and marker-based roentgen stereophotogrammetry methods in a typical clinical setting. J. Arthroplasty **24**(4), 594–606 (2009)
16. Jacobs, C.A., Christensen, C.P., Greenwald, A.S., McKellop, H.: Clinical performance of highly cross-linked polyethylenes in total hip arthroplasty. J. Bone Joint Surg. Am. **89**(12), 2779–2786 (2007)
17. Kang, L., Galvin, A.L., Brown, T.D., Fisher, J., Jin, Z.M.: Wear simulation of ultra-high molecular weight polyethylene hip implants by incorporating the effects of cross-shear and contact pressure. Proc. Inst. Mech. Eng. H **222**(7), 1049–1064 (2008)
18. Kang, L., Galvin, A.L., Brown, T.D., Jin, Z., Fisher, J.: Quantification of the effect of cross-shear on the wear of conventional and highly cross-linked UHMWPE. J. Biomech. **41**(2), 340–346 (2008)

19. Kang, L., Galvin, A.L., Fisher, J., Jin, Z.: Enhanced computational prediction of polyethylene wear in hip joints by incorporating cross-shear and contact pressure in additional to load and sliding distance: effect of head diameter. J. Biomech. **42**(7), 912–918 (2009)

20. Kaptein, B.L., Valstar, E.R., Spoor, C.W., Stoel, B.C., Rozing, P.M.: Model-based RSA of a femoral hip stem using surface and geometrical shape models. Clin. Orthop. Relat. Res. **448**, 92–97 (2006)

21. Kerner, J., Huiskes, R., van Lenthe, G.H., Weinans, H., van Rietbergen, B., Engh, C.A., Amis, A.A.: Correlation between pre-operative periprosthetic bone density and post-operative bone loss in THA can be explained by strain-adaptive remodelling. J. Biomech. **32**(7), 695–703 (1999)

22. Livingston, B.J., Chmell, M.J., Spector, M., Poss, R.: Complications of total hip arthroplasty associated with the use of an acetabular component with a Hylamer liner. J. Bone Joint Surg. Am. **79**(10), 1529–1538 (1997)

23. Malchau, H., Herberts, P., Eisler, T., Garellick, G., Soderman, P.: The Swedish total hip replacement register. J. Bone Joint Surg. Am. **84-A**(Suppl 2), 2–20 (2002)

24. Maxian, T.A., Brown, T.D., Pedersen, D.R., Callaghan, J.J.: The Frank Stinchfield Award. 3-Dimensional sliding/contact computational simulation of total hip wear. Clin. Orthop. Relat. Res. **333**, 41–50 (1996)

25. Maxian, T.A., Brown, T.D., Pedersen, D.R., McKellop, H.A., Lu, B., Callaghan, J.J.: Finite element analysis of acetabular wear. Validation, and backing and fixation effects. Clin. Orthop. Relat. Res. **344**, 111–117 (1997)

26. Ong, K.L., Rundell, S., Liepins, I., Laurent, R., Markel, D., Kurtz, S.M.: Biomechanical modeling of acetabular component polyethylene stresses, fracture risk, and wear rate following press-fit implantation. J. Orthop. Res. **27**(11), 1467–1472 (2009)

27. Pedersen, D.R., Brown, T.D., Hillis, S.L., Callaghan, J.J.: Prediction of long-term polyethylene wear in total hip arthroplasty, based on early wear measurements made using digital image analysis. J. Orthop. Res. **16**(5), 557–563 (1998)

28. Pedersen, D.R., Brown, T.D., Maxian, T.A., Callaghan, J.J.: Temporal and spatial distributions of directional counterface motion at the acetabular bearing surface in total hip arthroplasty. Iowa Orthop. J. **18**, 43–53 (1998)

29. Shaver, S.M., Brown, T.D., Hillis, S.L., Callaghan, J.J.: Digital edge-detection measurement of polyethylene wear after total hip arthroplasty. J. Bone Joint Surg. Am. **79**(5), 690–700 (1997)

30. Sochart, D.H.: Relationship of acetabular wear to osteolysis and loosening in total hip arthroplasty. Clin. Orthop. Relat. Res. **363**, 135–150 (1999)

31. Sochart, D.H.: True wear rates as predictors of risk for osteolysis after total hip arthroplasty. J. Bone Joint Surg. Am. **83-A**(8), 1277 (2001)

32. Stilling, M.: Polyethylene wear analysis. Experimental and clinical studies in total hip replacement. Acta Orthop. **80**(337), 1–74 (2009)

33. Stilling, M., Larsen, K., Andersen, N.T., Balle, K., Kold, S.X., Rahbek, O.: The final follow-up plain radiograph is sufficient for clinical evaluation of polyethylene wear in total hip arthroplasty. Acta Orthop. **81**(5), 570–578 (2010)

34. Stilling, M., Nielsen, K.A., Soballe, K., Rahbek, O.: Clinical comparison of polyethylene wear with zirconia or cobalt-chromium femoral heads. Clin. Orthop. Relat. Res. **467**(10), 2644–2650 (2009)

35. Stilling, M., Rahbek, O., Soballe, K.: Inferior survival of hydroxyapatite versus titanium-coated cups at 15 years. Clin. Orthop. Relat. Res. **467**(11), 2872–2879 (2009)

36. Stilling, M., Soballe, K., Andersen, N.T., Larsen, K., Rahbek, O.: Analysis of polyethylene wear in plain radiographs. The number of radiographs influences the results. Acta Orthop. **80**(6), 675–682 (2009)

37. Sychterz, C.J., Engh Jr., C.A., Swope, S.W., McNulty, D.E., Engh, C.A.: Analysis of prosthetic femoral heads retrieved at autopsy. Clin. Orthop. Relat. Res. **358**, 223–234 (1999)

38. Sychterz, C.J., Engh Jr., C.A., Yang, A., Engh, C.A.: Analysis of temporal wear patterns of porous-coated acetabular components: distinguishing between true wear and so-called bedding-in. J. Bone Joint Surg. Am. **81**(6), 821–830 (1999)

39. Valstar, E.R., de Jong, F.W., Vrooman, H.A., Rozing, P.M., Reiber, J.H.: Model-based roentgen stereophotogrammetry of orthopaedic implants. J. Biomech. **34**(6), 715–722 (2001)

40. Valstar, E.R., Vrooman, H.A., Toksvig-Larsen, S., Ryd, L., Nelissen, R.G.: Digital automated RSA compared to manually operated RSA. J. Biomech. **33**(12), 1593–1599 (2000)

41. van Lenthe, G.H., Willems, M.M., Verdonschot, N., Waal Malefijt, M.C., Huiskes, R.: Stemmed femoral knee prostheses: effects of prosthetic design and fixation on bone loss. Acta Orthop. Scand. **73**(6), 630–637 (2002)

42. Vrooman, H.A., Valstar, E.R., Brand, G.J., Admiraal, D.R., Rozing, P.M., Reiber, J.H.: Fast and accurate automated measurements in digitized stereophotogrammetric radiographs. J. Biomech. **31**(5), 491–498 (1998)

43. Wong, C., Gehrchen, P.M., Kiaer, T.: Can experimental data in humans verify the finite element-based bone remodeling algorithm? Spine (Phila Pa 1976) **33**(26), 2875–2880 (2008)

44. Wong, C., Gehrchen, P.M., Kiaer, T.: Can additional experimental data in humans verify the finite element-based bone remodeling algorithm? Open Spine J. **2**, 12–16 (2010)

45. Wroblewski, B.M.: Direction and rate of socket wear in Charnley low-friction arthroplasty. J. Bone Joint Surg. Br. **67**(5), 757–761 (1985)

46. Yamaguchi, M., Bauer, T.W., Hashimoto, Y.: Three-dimensional analysis of multiple wear vectors in retrieved acetabular cups. J. Bone Joint Surg. Am. **79**(10), 1539–1544 (1997)

47. Yamaguchi, M., Hashimoto, Y., Akisue, T., Bauer, T.W.: Polyethylene wear vector in vivo: a three-dimensional analysis using retrieved acetabular components and radiographs. J. Orthop. Res. **17**(5), 695–702 (1999)

Wear Analysis of Highly Cross-Linked Polyethylene in Total Hip Arthroplasty

13

Charles R. Bragdon, Michael Doerner, and Henrik Malchau

13.1
Introduction

Highly cross-linked polyethylene (HXLPE) has been introduced to decrease osteolysis secondary to polyethylene wear debris generation, thereby anticipating increasing long-term survivorship of the total hip arthroplasty (THA) [1–5]. There are a variety of types of highly cross-linked polyethylenes sold worldwide today. Due to different processing parameters, these vary on the degree of cross-linking; oxidative resistance; mechanical properties; and the presence of additives [6–11]. The electron-beam irradiated cross-linked and melted polyethylene was one of the first of the contemporary formulations to be introduced for clinical use at the end of 1999 and had been extensively evaluated in vitro in a series of pre-clinical tests [12–17]. One of the most clinically relevant wear tests for total hip arthroplasty evaluation has become the hip simulator and several of these have been developed for this purpose. Muratoglu et al. tested this formulation of highly cross-linked polyethylene up to 27 million cycles of simulated gait using the Boston hip stimulator [18]. That study documented both the reduction of secondary oxidation and improved wear of this form of highly cross-linked polyethylene by gravimetric analysis in comparison with results using conventional polyethylene total hip replacement components.

HXLPE is now one of the most widely utilized bearing surfaces for total hip arthroplasty (THA). A number of clinical studies have been initiated as different formulations of highly cross-linked polyethylene were developed and marketed. There are two general methods used for the clinical wear evaluation of bearings in THA. The most accurate method is radiostereometric analysis (RSA) [19–22]. While this method is the most technically challenging and currently can only be used in small groups of patients, it has been used extensively to provide early assurance that these new materials perform clinically as expected [23–28]. The other method utilizes semi-automated image analysis in order to determine the change in position of the center of the femoral head in relation to the center of the acetabular

C.R. Bragdon (✉), M. Doerner, and H. Malchau
The Harris Orthopaedic Laboratory, Massachusetts General Hospital, 55 Fruit Street, GRJ 1126, Boston, MA 02114, USA
e-mail: cbragdon@partners.org

K. Knahr (ed.), *Tribology in Total Hip Arthroplasty*,
DOI: 10.1007/978-3-642-19429-0_13, © 2011 EFORT

component over time in a series of pelvic radiographs. While there are a number of software programs that are used, the two most popular and well documented and validated are the Martell Hip Analysis Suite (Chicago University, Chicago Il) developed by Dr John Martell [29–33] and the PolyWear software (Australia) developed by Dr Peter Devane [34–36]. These early studies using these methodologies have provided insight into the early bedding-in period where the polyethylene creeps and plastically deforms, and the early femoral head penetration performance after the majority of bedding-in occurs [33,37–39]. The magnitude of bedding-in varies with different formulations of highly cross-linked polyethylene but in all instances is small and clinically irrelevant, resulting in femoral head penetrations in the range of 0.01–0.1 mm and is considered to continue at a very low rate after about 1 year in vivo. While there have been a few instances of reported polyethylene fractures due to malpositioning of the implants, there have been no serious concerns expressed in the use of any of the highly cross-linked polyethylenes currently used in THA.

The electron-beam irradiated cross-linked and melted polyethylene was developed in collaboration between researchers at Massachusetts General Hospital and the Massachusetts Institute of Technology. The first implantation of this new material in the USA was performed at MGH. At the time, both RSA studies and a larger clinical outcomes study were initiated. This study, initiated at the time that the new highly cross-linked polyethylene was made clinically available, is the longest term clinical follow-up series on this material. After the initial introductory period, continued long-term follow-up and performance evaluation is needed. The purpose of this study is to report on the 7–10 year clinical and radiographic outcomes of the first group of patients implanted with electron-beam irradiated cross-linked and melted polyethylene which was radiated to a dose of 10 mrads. This dose has been shown to provide the maximum reduction in wear in hip simulator studies [8,13,40–43].

13.2
Methods

All clinical data of patients of the Arthroplasty Service are collected in an IRB-approved local joint registry. For this study, the registry is queried periodically for data on primary total hip replacement patients implanted between January 1, 1999 and December 31, 2002. This patient group represents a minimum 7–10 year follow-up cohort and comprises 385 primary total hip replacements (355 patients) in which highly cross-linked polyethylene liners (Longevity or Durasul, Zimmer Inc.) with 22, 26, 28, or 32 mm diameter femoral heads were implanted. There were 186 females and 169 males with an average age of 61.7 years at surgery. The clinical measures used to evaluate these patients at each follow-up office visit were the Harris Hip Score, EQ-5D, SF-36 functional scores, and the UCLA activity score. In addition to conventional plain radiograph assessment, the Martell method was used to measure femoral head penetration over time.

For comparative purposes, a matched set of primary total hip replacement patients having conventional gamma sterilized in air polyethylene were identified and used as a Martell method wear measurement control group. Due to the varying number of patients in each femoral head size group, only a direct comparison between the penetration rates of patients having 26 and 28 mm diameter femoral heads with conventional or highly cross-linked polyethylene was performed.

In order to separate the early bedding-in process from the true steady-state penetration rate, penetration occurring between the 1 year film and each subsequent film was determined using the Martell Hip Analysis Suite software. We have developed two complimentary methods of summarizing the femoral head penetration data. (1) The penetration rate was defined as the slope of the linear regression line of the plot of the femoral head penetration occurring between the 1 year post-operative film and each subsequent film. This method uses all of the available data. Slopes were compared using the Zar test. (2) An average femoral head penetration rate was calculated using the measurement comparing the latest follow-up radiograph to the 1 year follow-up film. This method uses only one measurement for each patient and therefore spurious wear measurements have a greater effect on the results which is reflected in the standard deviation. A repeated-measures mixed model ANOVA was performed between the HXLPE and conventional steady-state wear rates for those patients with 26 or 28 mm femoral heads where a significant difference was defined at $p \leq 0.05$.

13.3
Results

The control group cohort consists of 141 hips (96 patients) with 26 mm femoral head size and 45 hips (33 patients) with 28 mm femoral head size coupled with conventional gamma sterilized in air polyethylene implanted by a single surgeon with a minimum of 7 years follow-up (range7–20 years). There was no significant difference between the wear of the two femoral head sizes ($p=0.43$), so the groups were combined to form a control cohort of 186 hips (129 patients). The group consists of 37 females and 92 males with an average age of surgery of 61.6 years.

In the highly cross-linked cohort, 172 THAs with femoral head diameters of 26 mm ($n=4$), 28 mm ($n=97$), and 32 mm ($n=71$) had appropriate radiographs for Martell Analysis. All of these patients had a minimum of 7 year radiographic follow-up. Currently, 38 patients are at a minimum 10 year follow-up. The minimum 10 year group comprises both 28 mm ($n=25$) and 32 mm ($n=13$) femoral head diameters.

Upon plain radiographic review, none of the HXLPE components showed radiographic loosening, failure or fracture. There were no osteolytic lesions visible around the acetabular or femoral components in either radiographic projection. Nor were any revisions performed for polyethylene wear or liner fracture.

Clinical outcome scores were obtained at each office visit. The latest score for each patient were tabulated in order to calculate an average score for each outcome measure. The average scores were: Harris Hip 88.1 ± 11.97, EQ-5D 74.0 ± 27.0, SF-36 physical activity scores 53.3 ± 8.4, SF-36 mental score 46.9 ± 11.1, and UCLA activity 6.4 ± 2.1 (Table 13.1).

13

Table 13.1 A summary of the average outcome scores at the latest follow-up of the THA patients with highly cross-linked polyethylene

Survey	N	Average±StDev
VAS pain	122	0.96±1.46
VAS satisfaction	120	0.8±1.66
EQ-5D Health Index	122	0.84±0.19
Harris Hip Score	136	88.28±14.46
SF-36 Physical Sum	113	47.3±10.77
SF-36 Mental Sum	113	54.48±8.52
UCLA activity	122	6.33±2.12
Womac pain	125	1.45±2.86

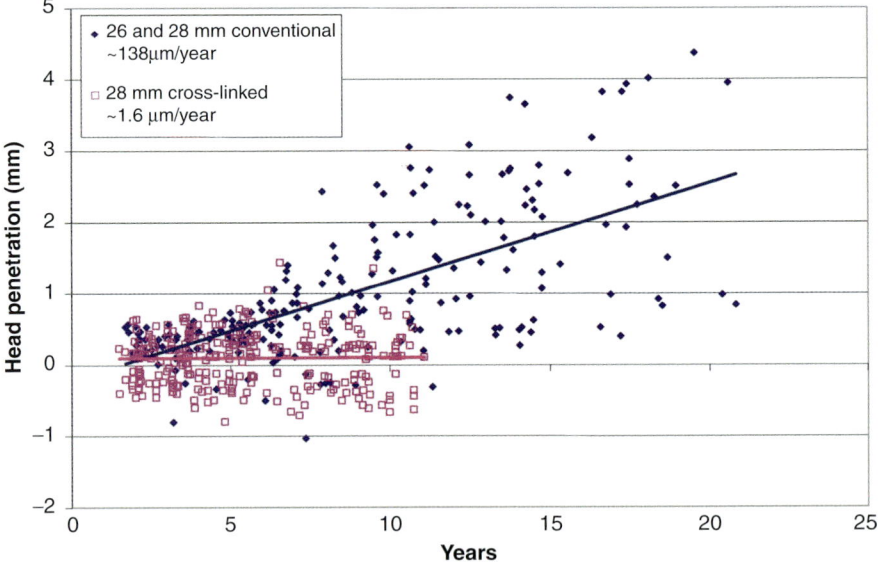

Fig. 13.1 The scatter plot of the penetration data from the conventional group having either 26 or 28 mm diameter femoral heads out to 20 years and the highly cross-linked group out to 11 years having 28 mm diameter femoral heads. The slope of the regression line is essentially zero and significantly lower than that of the conventional polyethylene group

The scatter plot of the conventional and highly cross-linked polyethylene groups is shown in Fig. 13.1. The steady-state wear of the conventional polyethylene patients increased with time and the slope of the regression line indicates a femoral head penetration rate of 0.138 mm/year (138 microns/year). The scatter of the control data also increases with time. In contrast, the femoral head penetration in the highly cross-linked polyethylene 28 mm femoral head diameter group did not increase over time and the slope of 0.0016 mm/year,

(1.6 microns/year) was not statistically discernable from a zero slope. The scatter of the cross-linked data did not increase with time. After repeated-measures mixed model ANOVA, the steady state wear rate of conventional polyethylene liners was significantly different than a slope of zero ($p<0.0001$). The wear of conventional polyethylene coupled with 26 or 28 mm femoral heads was significantly greater than that of the highly cross-linked liners ($p<0.0001$). The steady state wear rates of the highly cross-linked liners were not significantly different than a slope of zero ($p=0.54$).

Scatter plots were created for all patients representing a minimum 7 year follow-up as well as for the smaller group of patients with a minimum 10 year follow-up (Figs. 13.2 and 13.3). In neither case did the slope of the regression line differ from a zero slope and therefore there was no indication that the femoral head penetration increased with increasing time in vivo.

Finally, the average steady state femoral head penetration rate was calculated by using the latest follow-up radiograph and comparing it to the 1 year film for both the minimum 7 year follow-up and the minimum 10 year follow-up group for the 28 and 32 mm diameter femoral head cohorts (Table 13.2). There was no significant difference between the calculate average penetration rate between the 28 and 32 mm femoral head size groups at either a minimum 7 or 10 year follow-up. The calculated average penetration rates were extremely low. A graph of the penetration data against each UCLA activity score (Fig. 13.4) indicates that there is no correlation between activity and the magnitude of the measured femoral head penetration.

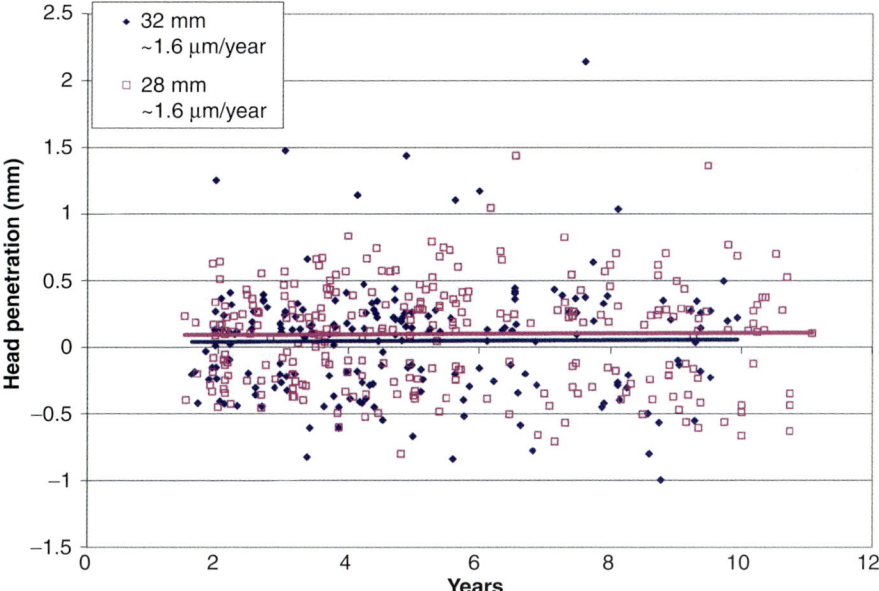

Fig. 13.2 A scatter plot of the highly cross-linked femoral head penetration data at a minimum of 7 years follow-up divided by femoral head size. The slope of each group is essentially zero and there is no correlation between femoral head penetration and time in vivo

13

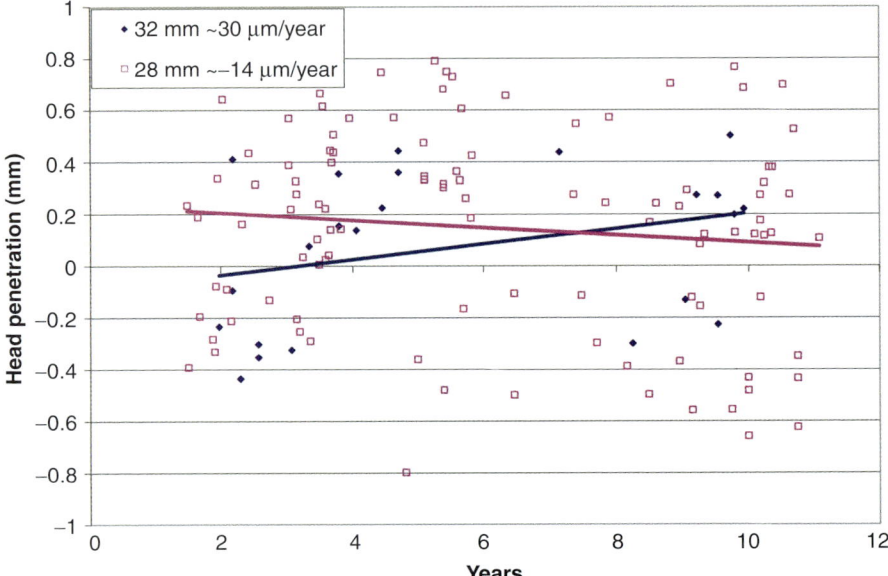

Fig. 13.3 A scatter plot of the highly cross-linked femoral head penetration data at a minimum of 10 years follow-up divided by femoral head size. While the slopes are slightly positive or negative, the slope of each group is essentially zero and there is no correlation between femoral head penetration and time in vivo

Table 13.2 The average steady state penetration rate of the patients having highly cross-linked THA at a minimum 7 and 10 year follow-up calculated by using the 1 year and the latest follow-up radiograph

Follow-up	Head size (mm) and number of patients	Average steady state rate (mm/year)
Minimum 7 years	28 ($n=97$)	0.007 ± 0.08
	32 ($n=71$)	0.004 ± 0.09
Minimum 10 years	28 ($n=25$)	0.012 ± 0.05
	32 ($n=13$)	0.027 ± 0.08

13.4
Discussion

Patients with implanted conventional gamma polyethylene, show significant wear of the material with femoral head penetration rates that are significantly different than zero (a slope of zero assumes no material wear) with both 26 and 28 mm femoral heads. Conversely, patients with highly cross-linked polyethylene display no measureable wear.

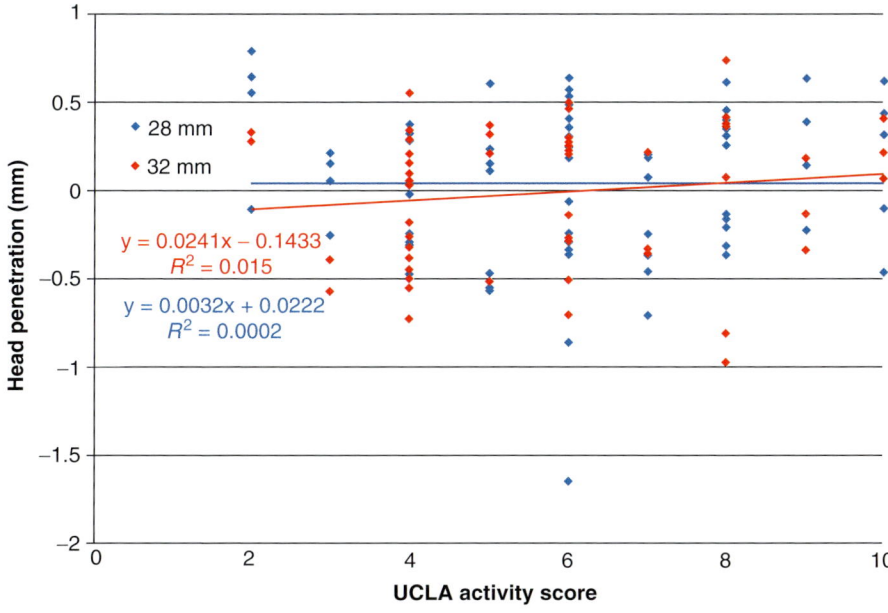

Fig. 13.4 There was no correlation between the amount of femoral head penetration and the UCLA activity score

HXLPE coupled with both 28 and 32 mm femoral head sizes had wear rates within the error detection of Hip Analysis Suite and were not significantly different from a wear rate of zero.

While one limitation of this chapter is the incomplete follow-up and femoral head penetration measurements on all patients, the results are consistent with previous reports on this patient cohort and there are no contradictory data to date that challenges the reported results. In addition, the rate of periprosthetic osteolysis around primary total hip arthroplasty using polyethylene as a bearing surface have been reported to range from 10% to 37% by 7–10 years after surgery. There is no evidence of osteolysis in any patients in this series and there have been no reports of definitive cases of osteolysis related to polyethylene debris in patients having this form of highly cross-linked polyethylene. Another limitation of this study may be the technique of measuring femoral head penetration from sequential AP radiographs. However, the Martell technique has been validated in many studies and is the most robust and developed software program for this purpose. When significant femoral head penetration due to wear occurs, such as with conventional polyethylene presented in this study, a steady increase in femoral head penetration, correlating with time in vivo, can be measured. The fact that sequential measurements spanning 7–10 years follow-up continue to show no correlation with time in vivo supports observations from retrieval studies of highly cross-linked acetabular liners which indicate minimal material has actually been removed from the articular surface.

This long-term clinical and radiographic follow-up study of patients receiving primary THR using highly cross-linked polyethylene liners represents the largest series and longest follow-up period for this new bearing material. The clinical follow-up results are similar to what would be expected in a primary THR patient population. The radiographic results are excellent with no signs of peri-prosthetic osteolysis. The wear results continue to indicate very low wear in vivo with no signs of changes over time.

Disclaimer Each author certifies that his or her institution has approved the human protocol for this investigation and that all investigations were conducted in conformity with ethical principles of research.

References

1. Harris, W.H.: Osteolysis and particle disease in hip replacement. Acta Orthop. Scand. **65**(1), 113–123 (1994)
2. Jasty, M., Haire, T., Tanzer, M.: Femoral osteolysis: a generic problem with cementless and cemented femoral components. Orthop. Trans. **15**, 758–759 (1991)
3. Jasty, M., et al.: Endosteal osteolysis around well-fixed porous-coated cementless femoral components. In: St. John, K.R. (ed.) Particulate Debris from Medical Implants. ASTM STP 1144, Philadelphia (1992)
4. Maloney, W.J., et al.: Fixation, polyethylene wear, and pelvic osteolysis in primary total hip replacement. Clin. Orthop. Relat. Res. **369**, 157–164 (1999)
5. Willert, H.G., Bertram, H., Buchhorn, G.H.: Osteolysis in alloarthroplasty of the hip. The role of ultra-high molecular weight polyethylene wear particles. Clin. Orthop. Relat. Res. **258**, 95–107 (1990)
6. Kurtz, S.M., et al.: Advances in the processing, sterilization, and crosslinking of ultra-high molecular weight polyethylene for total joint arthroplasty. Biomaterials **20**(18), 1659–1688 (1999)
7. Manley, T.R., Qayyum, M.M.: Crosslinked polyethylene at elevated temperatures. Polymer **12**, 177–181 (1971)
8. Jacobs, C.A., Christensen, C.P., Greenwald, A.S., McKellop, H.: Clinical performance of highly cross-linked polyethylenes in total hip arthroplasty. J Bone Joint Surg Am. **89**(12), 2779–2786 (2007)
9. Muratoglu, O.K., et al.: A comparison of five different types of highly crosslinked UHMWPEs: physical properties and wear behavior. In: Annual Meeting of Orthopaedic Research Society, Anaheim (1999)
10. Narkis, M., et al.: Structure and tensile behavior of irradiation- and peroxide-crosslinked polyethylenes. J. Macromol. Sci. **B26**(1), 37–58 (1987)
11. Oral, E., et al.: Alpha-tocopherol-doped irradiated UHMWPE for high fatigue resistance and low wear. Biomaterials **25**(24), 5515–5522 (2004)
12. Bragdon, C.R., O'Connor, D.O., Lowenstein, J.D., Jasty, M., Biggs, S.A., Harris, W.H.: A new pin-on-disk wear testing method for simulating wear of polyethylene on cobalt-chrome alloy in total hip arthroplasty. J Arthroplasty. **16**(5), 658–665 (2001)
13. Muratoglu, O.K., et al.: A novel method of cross-linking ultra-high-molecular-weight polyethylene to improve wear, reduce oxidation, and retain mechanical properties. Recipient of the 1999 HAP Paul Award. J. Arthroplasty **16**(2), 149–160 (2001)
14. Estok, D.M., et al.: The measurement of creep in ultrahigh molecular weight polyethylene: a comparison of conventional versus highly cross-linked polyethylene. J. Arthroplasty **20**(2), 239–243 (2005)

15. Bragdon, C.R., et al.: The importance of multidirectional motion on the wear of polyethylene. Proc. Inst. Mech. Eng., H: J. Eng. Med. **210**(3), 157–165 (1996)
16. Bragdon, C.R., et al.: Third-body wear testing of a highly cross-linked acetabular liner: the effect of large femoral head size. J. Arthroplasty **20**(3), 379–385 (2005)
17. Bragdon, C., et al.: Third-body wear of highly cross-linked polyethylene in a hip simulator. J. Arthroplasty **18**(5), 553–561 (2003)
18. Muratoglu, O., et al.: A Highly crosslinked, melted UHMWPE: expanded potential for total joint arthroplasty. In: Rieker, C., Oberholzer, S., Wyss, U. (eds.) World Tribology Forum in Arthroplasty, pp. 245–262. Hans Huber, Bern (2001)
19. Borlin, N., Thien, T., Karrholm, J.: The precision of radiostereometric measurements. Manual vs. digital measurements. J. Biomech. **35**(1), 69–79 (2002)
20. Bragdon, C.R., et al.: Experimental assessment of precision and accuracy of radiostereometric analysis for the determination of polyethylene wear in a total hip replacement model. J. Orthop. Res. **20**(4), 688–695 (2002)
21. Borlin, N., Rohrl, S.M., Bragdon, C.R.: RSA wear measurements with or without markers in total hip arthroplasty. J. Biomech. **39**(9), 1641–1650 (2006)
22. Bragdon, C.R., et al.: Standing versus supine radiographs in RSA evaluation of femoral head penetration. Clin. Orthop. Relat. Res. **448**, 46–51 (2006)
23. Digas, G., et al.: The Otto Aufranc Award. Highly cross-linked polyethylene in total hip arthroplasty: randomized evaluation of penetration rate in cemented and uncemented sockets using radiostereometric analysis. Clin. Orthop. Relat. Res. **429**, 6–16 (2004)
24. Rohrl, S., et al.: In vivo wear and migration of highly cross-linked polyethylene cups a radiostereometry analysis study. J. Arthroplasty **20**(4), 409–413 (2005)
25. Rohrl, S.M., et al.: Porous-coated cups fixed with screws: a 12-year clinical and radiostereometric follow-up study of 50 hips. Acta Orthop. **77**(3), 393–401 (2006)
26. Bragdon, C.R., et al.: Radiostereometric analysis comparison of wear of highly cross-linked polyethylene against 36- vs 28-mm femoral heads. J. Arthroplasty **22**(6 Suppl 2), 125–129 (2007)
27. Digas, G., et al.: 5-year experience of highly cross-linked polyethylene in cemented and uncemented sockets: two randomized studies using radiostereometric analysis. Acta Orthop. **78**(6), 746–754 (2007)
28. Karrholm, J., et al.: Five to 7 years experiences with highly cross-linked PE. In: SICOT TWC Abstract #19059. Hong Kong http://www.sicot.org/. 2008
29. Martell, J., et al.: Results of primary total hip reconstruction with the cementless Harris-Galante prosthesis: minimum five year results. Orthop. Trans. **15**, 750 (1991)
30. Martell, J.M., Berdia, S.: Determination of polyethylene wear in total hip replacements with use of digital radiographs. J. Bone Joint Surg. Am. **79**(11), 1635–1641 (1997)
31. Bragdon, C.R., et al.: A five year clinical comparison of the measurement of femoral head penetration in THR using RSA and the Martell method. In: 50th Annual Meeting Orthopaedic Research Society, San Francisco, (2004)
32. Bragdon, C.R., et al.: Comparison of femoral head penetration using RSA and the Martell method. Clin. Orthop. Relat. Res. **448**, 52–57 (2006)
33. Greene, M.E., et al.: A comparison of the Devane and Martell methods for in vivo measurement of polyethylene wear of patients with highly cross-linked polyethylene THR components. In: 54th Annual Meeting of the Orthopaedic Research Society, San Francisco (2008)
34. Devane, P.A., et al.: Measurement of polyethylene wear in metal-backed acetabular cups. I. Three-dimensional technique. Clin. Orthop. Relat. Res. **319**, 303–316 (1995)
35. Devane, P.A., et al.: Measurement of polyethylene wear in metal-backed acetabular cups. Clin. Orthop. Relat. Res. **319**, 317–326 (1995)
36. Devane, P.A., Horne, J.G.: Assessment of polyethylene wear in total hip replacement. Clin. Orthop. Relat. Res. **369**, 59–72 (1999)

37. Martell, J.M., Verner, J.J., Incavo, S.J.: Clinical performance of a highly cross-linked polyethylene at two years in total hip arthroplasty: a randomized prospective trial. J. Arthroplasty **18**(7 Suppl 1), 55–59 (2003)
38. Manning, D., et al.: In vivo wear comparison of traditional vs e-beam irradiated, post-irradiation melted, highly crosslinked polyethylene. J. Arthroplasty **20**(7), 880–886 (2005)
39. Bragdon, C.R., et al.: Steady state penetration rates of electron beam-irradiated, highly cross-linked polyethylene at an average 45-month follow-up. J. Arthroplasty **21**(7), 935–943 (2006)
40. McKellop, H., et al.: Development of an extremely wear resistant ultra-high molecular weight polyethylene for total hip replacements. J. Orthop. Res. **17**(2), 157–167 (1999)
41. McKellop, H., et al.: Effect of sterilization method and other modifications on the wear resistance or acetabular cups made of ultra-high molecular weight polyethylene. J. Bone Joint Surg. **82-A**(12), 1708–1725 (2000)
42. Muratoglu, O.K., et al.: Identification and quantification of radiation in UHMWPE. In: Annual Meeting of Orthopaedic Research Society, Anaheim (1999)
43. Muratoglu, O.K., et al.: Electron beam crosslinking of UHMWPE at room temperature, a candidate bearing material for total joint arthroplasty. In: 23rd Annual Meeting of the Society for Biomaterials, New Orleans (1997)

Rates of Osteolysis in Well-Functioning Alumina-on-Highly Cross-Linked Polyethylene Bearing Cementless THA in Patients Younger than Fifty with Femoral Head Osteonecrosis

14

Young-Hoo Kim, Yoowang Choi, and Jun-Shik Kim

14.1
Introduction

With early generation of cementless acetabular and femoral components of total hip arthroplasty (THA), durability of fixation has been excellent in the patients younger than 50 years of age out to 15 and 20 years [1–4]. Wear-induced osteolysis, however, is the dominant problem. Alternative bearing couples such as alumina ceramic-on-highly cross-linked polyethylene is attractive because of the potential for reduced wear and anticipated reduced osteolysis and loosening.

Ceramic bearings offer the advantage of improved lubrication, smoother surface finish, and improved resistance to scratching and are biologically inert compounds. Ceramic femoral heads therefore have substantial tribologic advantages over metal femoral heads and result in much lower wear and osteolysis rates [5]. Also, highly cross-linked ultra-high molecular weight polyethylene has been shown to markedly reduce wear in clinical studies [6, 7]. Furthermore, alumina-on-highly cross-linked polyethylene bearing couple has potential for further reduction of polyethylene wear and osteolysis.

The purpose of the current study was to evaluate the clinical and radiographic outcomes of THAs using alumina-on-highly cross-linked polyethylene bearing couple in 71 patients younger than 50 with femoral head osteonecrosis. Additionally, we determined the incidence of polyethylene wear and osteolysis using radiographs and computer tomographic scans.

Y.-H. Kim (✉)
The Joint Replacement Center of Korea, at Ewha Womans University MokDong Hospital,
911-1, MokDong, YangCheon-Gu, Seoul, Korea
e-mail: younghookim@ewha.ac.kr

Y. Choi and J.-S. Kim
The Joint Replacement Center of Korea, Ewha Womans University School of Medicine,
Seoul, Korea

K. Knahr (ed.), *Tribology in Total Hip Arthroplasty*,
DOI: 10.1007/978-3-642-19429-0_14, © 2011 EFORT

14.2
Materials and Methods

14.2.1
Demographics

From February 2000 to May 2002, the senior author performed 79 consecutive cementless THAs using alumina-on-highly cross-linked polyethylene (Marathon™, DePuy, Warsaw, IN) in 76 patients (three patients had bilateral THAs). The study was approved by our institutional review board and all patients provided informed consent. Five patients were lost to follow-up (before 1 year), leaving 71 patients (73 hips), who comprise the series of this study.

There were 48 men, (50 hips) and 23 women (23 hips). The average age at the time of the index arthroplasty was 45.5 years (range, 20–50 years). The average weight of the patients was 67.5 kg (range of 47–87 kg). The average height was 164.5 cm (range, 147–183 cm) and the average body-mass index was 24.8 kg/m^2 (range, 18.8–30.5 kg/m). All hips with osteonecrosis of femoral head had Ficat and Arlet stage III or IV [8]. The presumed cause of osteonecrosis was ethanol abuse in 50 patients (70.4%), idiopathic in 18 patients (25.4%), and steroid use in 3 patients (4.2%). The average follow-up was 8.5 years (range, 7–9 years).

14.2.2
Prosthesis

A cementless Duraloc 100 or 1200 series acetabular component with a highly cross-linked polyethylene liner of inner diameter of 28 mm was used in all hips. All patients received an Immediate Postoperative Stability (IPS; DePuy, Leeds, United Kingdom) cementless femoral component with a 28 mm alumina forte femoral head. The IPS femoral component is an anatomical metaphyseal-fitting titanium stem with a polished and tapered distal stem, designed to provide fixation in the metaphysis only, thereby avoiding metal-to-bone contact below this point. The proximal 30% of stem was porous-coated with sintered titanium beads, with a mean pore size of 250 μm to which a hydroxy apatite coating was applied to a thickness of 30 μm.

14.2.3
Surgical Procedure

All operations were performed through a posterolateral approach. The femoral component was inserted with a press-fit technique. The largest broach that would fill the metaphysis and leave little cancellous bone remaining was used. The acetabular component was fixed with a press-fit only without using a screw in the 65 hips and one or two screws were inserted for additional fixation in the remaining eight hips.

The patients were allowed to stand on the second postoperative day and they progressed to full weight-bearing with crutches as tolerated.

14.2.4
Clinical Evaluation

Clinical follow-up was performed at 3 months and 1 year, and yearly thereafter. Harris hip scores were determined before surgery and at each follow-up examination [9]. Patients subjectively evaluated thigh pain on a 10-point visual analog scale ($0=$no pain; $10=$severe pain). The level of activity of the patients after the THA was assessed with the activity score of Tegner and Lysholm [10]. The activity grading scale, with which work and sports activities are graded numerically, was used as a complement to the functional score. The patients were given a score, according to the activities in which they engaged in daily life, ranging from 0 points for a hip-related disability to 10 points for participation in competitive sports at a national level.

14.2.5
Radiographic Evaluation

Radiographic follow-up was performed at 3 months and 1 year and yearly thereafter. A supine anteroposterior radiograph of the pelvis with both hips in neutral rotation and $0°$ of abduction was made for every patient. Consistent patient positioning was ensured with the use of x-ray frame. This frame is constructed so that it can be placed at the end of a standard x-ray table. Plastic polypropylene orthosis are secured to a plastic backboard through a vertical slot. A wing nut allows adjustment for various limb lengths. Rotation and abduction remain constant. Cross-table lateral radiographs were also made of each hip.

Femoral bone type was determined in preoperative radiographs using Dorr's classification [11]. The adequacy of the intramedullary fill by the stem was recorded as satisfactory when the stem filled >80% of the proximal part of the canal in the coronal plane and >70% in the sagittal plane, according to a previously described method [12]. The component was considered to be undersized if less of the canal was filled in either or both planes.

Definite loosening of the femoral component was defined if there was a progressive axial subsidence of >2 mm or varus or valgus shift [13]. A femoral component was considered to be possibly loose when there was a complete radiolucent line surrounding the entire porous-coated surface on both the anteroposterior and lateral radiograph [13].

Anteversion of the acetabular component was measured on the true lateral radiographs of the hip as the angle between a horizontal line and a second line marking the plane of the socket. To measure cup abduction, a line that joined the inferior margins of the two acetabular teardrops was drawn on the anteroposterior pelvic radiograph. The intersection of that line marking the plane of opening of the socket determined the angle of abduction.

Definite loosening of the acetabular component was diagnosed when there was a change in the position of the component (>2 mm vertically and/or medially or laterally) or a continuous radiolucent line wider than 2 mm on both the anteroposterior and the lateral radiograph [13].

14

A vertical change in the position of the cup was measured between the inferior margin of the cup and the inferior margin of the ipsilateral teardrop [14], and a horizontal change was measured between the Köhler line (ilioischial line) and the center of the outer shell of the acetabular component [15].

14.2.6
Radiographic Evaluation of Polyethylene Wear

Penetration of the polyethylene liner was measured, with use of a software program (Auto CAD, Release 13; Autodesk, Sauslito, California [16]), by on observer who was blinded to the clinical results. The observer made three measurements in each radiograph, and the intraobserver error was±0.021 mm. A scan-Maker 9600 X L flat-bed imaging scanner (Microtek, Carson, California) digitized the anteroposterior radiograph of the pelvis as two-dimensional gray-scale arrays of twelve-bit (256 gray level) integers. The scanning resolution was 600 psi (pixels per square inch). Penetration of the head into the liner, was determined at annual intervals from anteroposterior pelvic radiographs. The amount of penetration on radiographs made 3 months postoperatively was considered to be the "zero position".

14.2.7
Radiographic Evaluation of Osteolysis

The presence and locations of areas of osteolysis in the acetabulum were recorded in the anteroposterior and lateral radiographs according to the system of DeLee and Charnley [17], and those in the femur were recorded also in the anteroposterior and lateral radiographs according to the system of Gruen et al. [18]. The length and width of osteolytic lesions were measured, and the area was expressed square centimeters.

14.2.8
Computer Tomographic Evaluation of Osteolysis

Although radiologic evaluation of osteolysis is a direct measurement, the current methodology is insensitive and subject to operator error. A more sensitive computerized tomography image sets provide three-dimensional data but the beam hardening artifacts from the prosthesis itself make these images difficult to interpret and use. To address the beam hardening artifacts, as well as to measure the volume of lytic lesions, we developed an algorithm to diminish the effect of beam hardening artifacts. We then developed a segmentation algorithm to segment the lytic lesions from image data and to measure their volumes. Computer tomographic images were acquired using Siemens AG (Munich, Germany) with 1 mm collimation, a pitch of 1.5, and a 14–22 cm field of view. The raw data was reconstructed for 1 mm slices. The area within 5 cm from the prosthesis-bone interface in all directions was evaluated. The volume of osteolysis was calculated by Virtual Scopics (Rochester, New York).

14.2.9
Statistical Methods

Survivorship analysis [19] was performed with the Kaplan–Meier method with revision for any reason as one end point and revision due to mechanical failure (clinical and radiographic evidence of aseptic loosening) at the time of follow-up as the other end point. We determined differences in continuous variances (Harris hip score, range of motion, body mass index) between preoperative and postoperative results using Student's paired t-test, and in categorical variances (details of functional evaluation and deformity according to the Harris hip score) and limb length between preoperative and postoperative evaluations using chi square test. Univariate regression analysis was used to evaluate the relationship, if any between osteolysis and the variables of age, gender, weight, diagnosis, the duration of follow-up, and acetabular inclination and anteversion. The level of significance was set at $P<0.05$.

14.3
Results

14.3.1
Clinical Outcome

Preoperative Harris hip score was improved significantly ($P=0.001$). The mean preoperative Harris hip score was 50.6 points (range, 27–55 points). The mean hip score at the final follow-up was 96 points (range, 85–100 points).

Preoperative functional activity was improved significantly ($P=0.001$) at the latest follow-up. The ability to put on footwear, and to cut toenails and to use stairs and public transportation was markedly improved at the latest follow-up.

Activity level of patients was improved very much after the operation. Many patients were quite active despite our admonitions to avoid activities involving high impact after the THA. All but three patients had an activity score of 5 or 6 points [10] after the THA, indicating participation in strenuous farm work (a score of 5 points) or playing recreational sports such as tennis (a scores of 6 points). The three patients with low back pain had a score of 3 points.

14.3.2
Radiographic Results

No hip had aseptic loosening of any acetabular or femoral component. All stems had a satisfactory canal fill on radiographs and all hips had Dorr type A or type B bones. The mean inclination of and anteversion of acetabular component was 42.3° (range, 39°–45°) and 22° (range, 19°–24°), respectively. All acetabular and femoral components were fixed by bone ingrowth. Calcar rounding off was observed in all hips but no hip had stress shielding related proximal femoral bone resorption.

No hip had femoral head ceramic fracture.

14

14.3.3
Radiographic Measurement of Penetration of Polyethylene

The mean amount of highly cross-linked polyethylene linear penetration was 0.05 ± 0.02 mm per year (range, 0.02–0.08 mm per year). At the latest follow-up examination, no hip was an outlier for the so-called osteolysis threshold of 0.10 mm per year [20, 21] with all liners having a penetration rate at or below this level. With the numbers available, univariate regression analysis did not demonstrate that age, gender weight, activity, cup inclination, or cup anterversion had any influence on polyethylene liner penetration.

14.3.4
Radiographic and Computer Tomographic Results of Osteolysis

Radiograph and computer tomographic scans (Fig. 14.1) demonstrated that no acetabular or femoral osteolysis was detected in any hip at the latest follow-up

14.3.5
Complications

One hip (1%) was dislocated 5 days after the operation and was treated successfully with a closed reduction and an abduction brace for 3 months. No further dislocation was observed in this hip until the final follow-up examination.

14.3.6
Revisions and Survivorship

No hip had a revision or aseptic loosening of acetabular and/or femoral component. Kaplan–Meier survival analysis, with revision as the endpoint for failure, showed that the rate of survival of both acetabular femoral components at 8 years was 100% (95% CI, 98–100).

14.4
Discussion

At a 8.5-year follow-up, the young patients in this series have performed well clinically and radiographically. These results are consistent with those in other studies of ceramic-on-highly cross-linked or conventional polyethylene [22, 23]. In one report on 56 THAs in young, active patient population, no patient had radiographic evidence of osteolysis and no patient had been revised for mechanical loosening or wear at an average of 30 months [22]. In another report on 100 patients younger than 50 years of age, no patient had radiographic

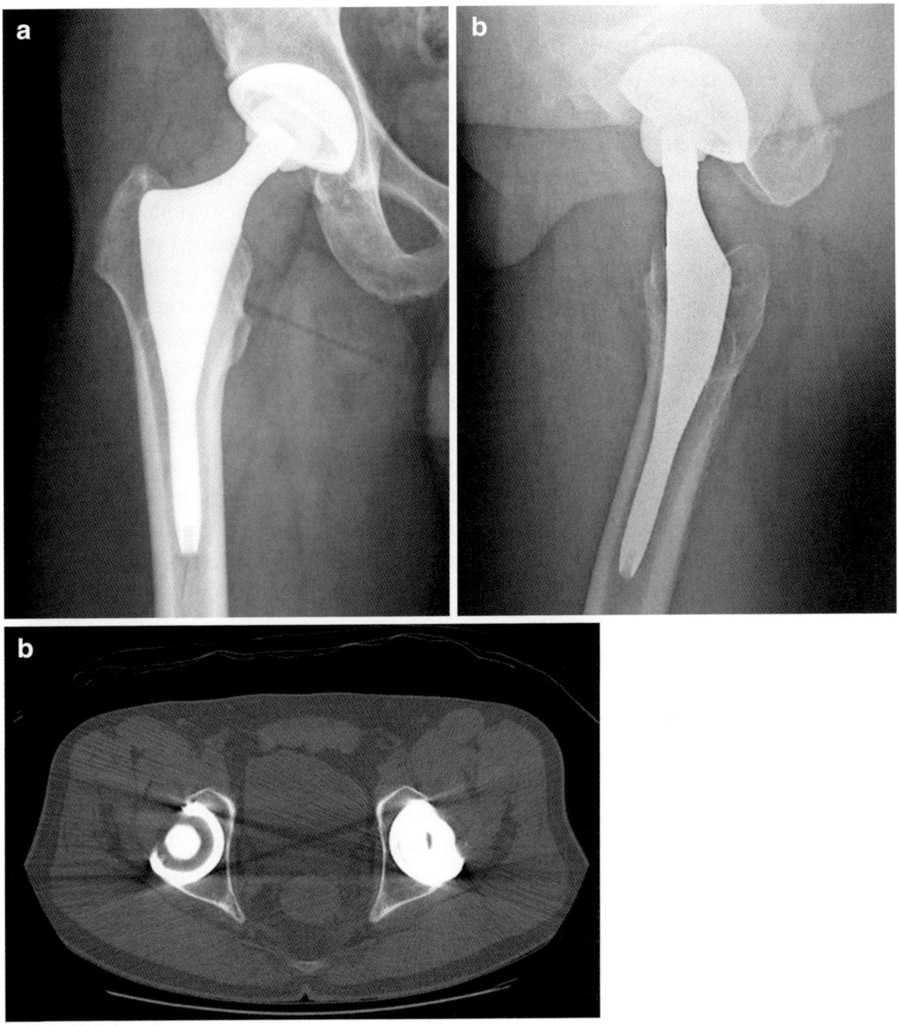

Fig. 14.1 Radiographic and computer tomographic scanning evaluation of osteolysis of right hip of a 43-year-old man with osteonecrosis of right hip. (**a**): An anteroposterior and lateral radiographs of right hip made 7 years after surgery demonstrated that Duraloc 1200 series cementless acetabular component is fixed in a satisfactory position by bone ingrowth. There in no radiolucent line or osteolysis around the acetabular or femoral components in both hips. Grade 3 bone loss is observed in the calcar region of right hip. (**b**): Computer tomographic scanning of right hip taken 7 years after the surgery reveals no evidence of osteolysis around the acetabular or femoral components in right hip

evidence of osteolysis and no patient had been revised for mechanical loosening or wear at an average of 5.6 years [23]. In the other report on 64 total hip prostheses in 56 patients who had ceramic-on-conventional polyethylene bearing, five patients (8%) had revision, but no patient had radiographic evidence of osteolysis at an average of 18 years [24]. They suggest

14

that the 8% revision rate of an average of 18 years in their study is relatively low, despite the use of what is now considered an inferior stem design (Charnley-Müller), first generation cementing technique, and a head size (32 mm) associated with greater volumetric wear.

Data on the outcome of highly cross-linked polyethylene and its effect on the prevention of osteolysis are limited. Results of studies out to 5 years have been reported for many of the commonly used highly cross-linked polyethylenes. In one study with a minimum 5-year follow-up using manual radiographic techniques to measure wear, the highly cross-linked polyethylene linear wear rate was 0.029 mm per year versus 0.065 mm per year for conventional polyethylene [25]. Using edge-detection techniques at 4 years, the rate for highly cross-linked polyethylene linear wear was 0.007 mm per year versus 0.174 mm per year for conventional polyethylene [26, 27]. Bitsch et al. [7] reported after a mean duration of follow-up of 5.8 years (range, 5.0–7.7 years) that the average femoral head penetration was 0.031 mm/year (range, 0.04–0.196 mm/year) in the hips with a highly cross-linked polyethylene liner and 0.104 mm/year (range 0.04–0.196 mm/year) in the hips with an Enduron polyethylene liner (DePuy). The wear rate (femoral head penetration) in the highly cross-linked polyethylene group was 71% lower than that in the Enduron group ($p=0.003$). Osteolysis was not observed in any of hips with a highly cross-linked polyethylene liner. Engh et al. [6]. reported a reduction in the mean wear rate of 95% for highly cross-linked polyethylene liners compared with Enduron liners (0.01 ± 0.12 mm/year) compared with 0.19 ± 0.12 mm/year).

Our penetration data for highly cross-linked polyethylene liners show an approximately 70% reduction in linear penetration when compared with the penetration rates for conventional polyethylene liners as reported in studies by several authors [14, 16, 21, 27]. In our series, the penetration rate of highly cross-linked polyethylene was high compared to other series. A bedding-in period of the polyethylene has been incorporated into our steady-state penetration rate. We believe that relatively higher penetration rate of highly cross-linked polyethylene in our series is related to young patient with osteonecrosis of femoral head. However, the penetration rate was still below the so-called osteolysis threshold [20, 21] (0.06 ± 0.03 mm/year). No hip in our series had detectable acetabular or femoral osteolysis.

Because of the well-recognized inability of plain x-rays to determine the presence or absence of osteolysis with high accuracy, it has been recommended to use a computer tomography analysis to determine presence of osteolysis with high accuracy. In a study using computer tomography analysis at nearly 6 years follow-up, possible osteolysis was observed in 8% (three of 36 hips) of THAs with highly cross-linked polyethylene, compared with 28% (11 of 40 hips) of THAs with gas plasma-sterilized polyethylene inserts; however, the absence of preoperative computer tomography scans makes the interpretation of these results less conclusive [28]. Furthermore, none of the patients in their study demonstrated evidence of loosening or pain, suggesting that osteolysis (at the early period of follow-up) may not necessarily correlate with implant failure. In the current series, we were not able to detect osteolysis in any of hips. Although no hip had an evidence of osteolysis in this series, the short duration of the follow-up cannot be accepted as establishing an absence of osteolysis.

First-generation highly cross-linked polyethylenes have documented reductions in fatigue, tensile, and toughness properties [28]. Polyethylene fractures have been associated with malpositioned acetabular components; edge loading has produced high contact stresses in the locking mechanism. Use of large heads with thin highly cross-linked polyethylene liners is a concern in cases with malpositioned acetabular components because fracture of the highly cross-linked polyethylene insert has been observed [29–31]. In the current series, no hip had polyethylene liner fracture. We believe that the use of adequate thickness of acetabular polyethylene liner (minimum 8 mm, average 9.4 mm, range 8–10.7 mm) and satisfactory position of acetabular component led to absence of polyethylene liner fracture.

There has been some concern that smaller wear particles are produced with highly cross-linked polyethylene than with conventional polyethylene [32], leading to a higher functional biological activity [33]. However, in our short-term data, no hip had an evidence of acetabular or femoral osteolysis. Therefore, the data here were compatible with no difference in the functional biologic activity between the highly cross-linked polyethylene and the conventional polyethylene. The longer term follow-up is mandatory to prove this biologic activity of highly cross-linked polyethylene.

In the current study, the absence of ceramic head fracture contrasts with the results of other studies on the use of contemporary designs with alumina femoral head [34–36]. We attribute the absence of ceramic femoral head fracture in our series to careful intraoperative handling of the ceramic head and the surgeon ensuring taper mating surfaces of the femoral component and the ceramic femoral head were aligned perfectly and remained debris-free during impaction.

There are some limitations to this study. First, although the follow-up period is too short to be conclusive, we observed a very low wear-rate of polyethylene. Second, the relatively high early dropout rate in this study may jeopardize the significance of the study. Finally, although this paper only reported on nonobese patients, all the patients in our series were young and active.

In conclusion, this current generation of anatomic tapered cementless femoral component with alumina-on-highly cross-linked polyethylene bearing couples is functioning well with no osteolysis at a 7-year minimum and average of 8.5-year follow-up in this series of young patients with osteonecrosis of the femoral head. While the long-term prevalence of ceramic head and acetabular highly cross-linked polyethylene fracture remains unknown, the short-term data are promising. Because of the anticipated reduction in wear long-term, alumina-on-highly cross linked polyethylene bearings are recommended for young patients requiring THA as well as those who desire to return to high demand athletics.

References

1. Della Valle, C.J., Berger, R.A., Shott, S., Rosenberg, A.G., Jacobs, J.J., Quigley, L., Galante, J.O.: Primary total hip arthroplasty with a porous-coated acetabular component: a concise follow-up of a previous report. J. Bone Joint Surg. Am. **86**, 1217 (2004)
2. Duffy, G.P., Prpa, B., Rowland, C.M., Berry, D.J.: Primary uncemented Harris-Galante acetabular components in patients 50 years old or younger: Results at 10 to 12 years. Clin. Orthop. Relat. Res. **427**, 157 (2004)
3. Gaffey, J.L., Callaghan, J.J., Pedersen, D.R., Goetz, D.D., Sullivan, P.M., Johnston, R.C.: Cementless acetabular fixation at fifteen years. A comparison with the same surgeon's results following acetabular fixation with cement. J. Bone Joint Surg. Am. **86**, 257 (2004)

14

4. Chen, C.J., Xenos, J.S., McAuley, J.P., Young, A., Engh Sr., C.A.: Second generation porous-coated cementless total hip arthroplasties have high survival. Clin. Orthop. Relat. Res. **451**, 121 (2006)

5. Jazrawi, L.M., Kummer, F.J., DiCesare, P.E.: Alternative bearing surfaces for total joint arthroplasty. J. Am. Acad. Orthop. Surg. **6**, 198 (1998)

6. Engh Jr., C.A., Stepniewski, A.S., Ginn, S.D., Beykirch, S.E., Sychterz-Terefenko, C.J., Hopper Jr., R.H., Engh, C.A.: A randomized prospective evaluation of outcomes after total hip arthroplasty using cross-linked marathon and non-cross-linked Enduron polyethylene liners. J. Arthroplasty **21**(6 suppl 2), 17 (2006)

7. Bitsch, R.G., Loidolt, T., Heisel, C., Ball, S., Schmalzried, T.P.: Reduction of osteolysis with use of Marathon cross-linked polyethylene. A concise follow-up, at a minimum of five years, of a previous report. J. Bone Joint Surg. Am. **90**, 1487 (2008)

8. Ficat, R.P., Arlet, J.: Treatment of bone ischemia and necrosis. In: Hungerford, D.S. (ed.) Ischemia and Necrosis of Bone, p. 171. Williams & Wilkins, Baltimore (1980)

9. Harris, W.H.: Traumatic arthritis of the hip after dislocation and acetabular fractures: treatment by mold arthroplasty. An end-result study using a new method of result evaluation. J. Bone Joint Surg. Am. **51**, 737 (1969)

10. Tegner, Y., Lysholm, J.: Rating systems in the evaluation of knee ligament injuries. Clin. Orthop. Relat. Res. **198**, 43 (1985)

11. Dorr, L.D.: Total hip replacement using APR system. Tech. Orthop. **1**, 22–34 (1986)

12. Kim, Y.-H., Kim, V.E.: Uncemented porous-coated anatomic total hip replacement. Results at six years in a consecutive series. J. Bone Joint Surg. Br. **75**, 6 (1993)

13. Kim Y-H Kim J-S, OhS-H, Kim, J.-M.: Comparison of porous-coated titanium femoral stem with and without hydroxyapatite coating. J. Bone Joint Surg. Am. **85**, 1682 (2003)

14. Sychterz, C.J., Engh Jr., C.A., Shah, N., Engh Sr., C.A.: Radiographic evaluation of penetration by the femoral head into the polyethylene liner over time. J. Bone Joint Surg. Am. **79**, 1040 (1997)

15. Sutherland, C.J., Wilde, A.H., Borden, L.S., Marks, K.E.: A ten-year follow-up of one hundred consecutive Müller curved-stem total hip-replacement arthroplasties. J. Bone Joint Surg. Am. **64**, 970 (1982)

16. Kim, Y.-H., Kim, J.-S., Chi, S.-H.: A comparison of polyethylene wear in hips with cobalt-chrome or zirconia heads. A prospective, randomized study. J. Bone Joint Surg. Br. **83**, 742 (2001)

17. DeLee, J.G., Charnley, J.: Radiological demarcation of cemented sockets in total hip replacement. Clin. Orthop. Relat. Res. **121**, 20 (1976)

18. Gruen, T.A., McNeice, G.M., Amstutz, H.C.: "Modes of failure" of cemented stem-type femoral components: a radiographic analysis of loosening. Clin. Orthop. Relat. Res. **141**, 17 (1979)

19. Kaplan, E.L., Meier, P.: Nonparametric estimation from incomplete observation. J. Am. Stat. Assoc. **53**, 457 (1958)

20. Wan, Z., Dorr, L.D.: Natural history of femoral focal osteolysis with proximal ingrowth smooth stem implant. J. Arthroplasty **11**, 718 (1996)

21. Dowd, J.E., Sychterz, C.J., Young, A.M., Engh, C.A.: Characterization of long-term femoral-head-penetration rates. Association with and prediction of osteolysis. J. Bone Joint Surg. Am. **82**, 1102 (2000)

22. Garvin, K.L., Hartman, C.W., Mangla, J., Murdoch, N., Martell, J.M.: Wear analysis in THA utilizing oxidized zirconium and crosslinked polyethylene. Clin. Orthop. Relat. Res. **467**, 141 (2009)

23. Kim, Y.-H., Kim, J.-S., Choi, Y.-W., Kwon, O.-R.: Intermediated results of simultaneous alumina-on-alumina bearing and alumina-on-highly cross-linked polyethylene bearing total hip arthroplasties. J Arhtroplasty **24**, 885 (2009)

24. Urban, J.A., Garvin, K.L., Boese, C.K., Bryson, L., Pedersen, D.R., Callaghan, J.J., Miller, R.K.: Ceramic-on-polyethylene bearing surfaces in total hip arthroplasty. Seventeen to twenty-one-year results. J. Bone Joint Surg. Am. **83**, 1688 (2001)

25. Dorr, L.D., Wan, Z., Shahrdar, C., Sirianni, L., Boutary, M., Yun, A.: Clinical performance of a Durasul highly cross-linked polyethylene acetabular liner for total hip arthroplasty at five years. J. Bone Joint Surg. Am. **87**, 1816 (2005)

26. Bragdon, C.R., Barrett, S., Martell, J.M., Greene, M.E., Malchau, H., Harris, W.H.: Steady-state penetration rates of electron beam-irradiated, highly cross-linked polyethylene at an average 45-month follow-up. J. Arhtroplasty **21**, 935 (2006)

27. Manning, D.W., Chiang, P.P., Martell, J.M., Galante, J.O., Harris, W.H.: In vivo comparative wear study of traditional and highly cross-linked polyethylene in total hip arthroplasty. J. Arhtroplasty **20**, 880 (2005)

28. Bradford, L., Baker, D., Ries, M.D., Pruitt, L.A.: Fatigue crack propagation resistance of highly crosslinked polyethylene. Clin. Orthop. Relat. Res. **429**, 68 (2004)

29. Tower, S.S., Currier, J.H., Currier, B.H., Lyford, K.A., Van Citters, D.W., Mayor, M.B.: Rim cracking of the cross-linked longevity polyethylene acetabular liner after total hip arthroplasty. J. Bone Joint Surg. Am. **89**, 2212 (2007)

30. Crowninshield, R.D., Maloney, W.J., Wentz, D.H., Humphrey, S.M., Blanchard, C.R.: Biomechanics of large femoral heads: what they do and don't do. Clin. Orthop. Relat. Res. **429**, 102 (2004)

31. Ries, M.D.: Highly cross-linked polyethylene: the debate is over. In opposition. J. Arhtroplasty **20**(4 supple 2), 59 (2005)

32. Fisher, J., McEwen, H.M., Tipper, J.L., Galvin, A.L., Ingram, J., Kamali, A., Stone, M.H., Ingham, E.: Wear, debris, and biologic activity of cross-linked polyethylene in the knee: benefits and potential concerns. Clin. Orthop. Relat. Res. **428**, 114 (2004)

33. Endo, M.M., Tipper, J.H., Barton, D.C., Stone, M.H.: Comparison of wear, wear debris and functional biological activity of moderately crosslinked and non-crosslinked polyethylenes in hip prostheses. Proc. Inst. Mech. Eng. H. **210**, 111 (2002)

34. D'Antonio, J., Capello, W., Manley, M., Bierbaum, B.: New experience with alumina-on-alumina ceramic bearings for total hip arthroplasty. J. Arthroplasty **17**, 390 (2002)

35. Capello, W.N., D'Antonio, J.A., Feinberg, J.R., Manley, M.T., Naughton, M.: Ceramic-on-ceramic total hip arthroplasty: update. J. Arthroplasty **23**(Suppl 1), 39 (2008)

36. Yoo, J.J., Kim, Y.M., Yoon, K.S., Koo, K.H., Song, W.S., Kim, H.J.: Alumina-on-alumina total hip arthroplasty. A five-year minimum follow-up study. J. Bone Joint Surg. Am. **87**, 530 (2005)

Osteolysis and Aseptic Loosening: Cellular Events Near the Implant

15

Gema Vallés, Eduardo García-Cimbrelo, and Nuria Vilaboa

15.1
Introduction

Total joint arthroplasty represents the most significant procedure achieved in orthopedic surgery to improve the quality of life of patients suffering from several end-stage joint diseases, mainly osteoarthritis. Among them, total hip replacement (THR) is one of the most common interventions, and it is recognized as the operation of the century [1]. The demand for such medical device implants is expected to rise in coming years because of the increase in life expectancy, and also because joint problems are gradually detected in younger and more active patients. The success achieved with hip implants and advances in femoral and acetabulum anchorage using bioactive and porous coatings or newly developed cements have improved implant durability, with remarkable functional benefits to the patients. Unfortunately, the solution offered by hip prostheses is limited in time. Long-term effects of debris generated from unavoidable and continuous wear due to the articulating motion at the bearing surfaces of the prosthetic hip and micromotion in non-articulating interfaces may cause osteolysis, leading to instability and aseptic loosening [2–4]. Nowadays, aseptic loosening is the most common reason for failure and revision surgery worldwide. Additionally, patients undergoing this type of revision have poorer functional outcomes and higher complication rates than patients undergoing primary arthroplasty [5].

N. Vilaboa (✉) and G. Vallés
Hospital Universitario La Paz-IdiPAZ, Paseo de la Castellana 261, 28046
Madrid, Spain and
Centro de Investigación Biomédica en Red de. Bioingeniería Biomateriales y
Nanomedicina, CIBER-BBN, Madrid, Spain
e-mail: nvilaboa.hulp@salud.madrid.org

E. García-Cimbrelo
Centro de Investigación Biomédica en Red de. Bioingeniería Biomateriales y
Nanomedicina, CIBER-BBN, Madrid, Spain and
Departamento de Cirugía Ortopédica y Traumatología, Hospital Universitario La Paz,
Paseo de la Castellana 261, 28046 Madrid, Spain

K. Knahr (ed.), *Tribology in Total Hip Arthroplasty*,
DOI: 10.1007/978-3-642-19429-0_15, © 2011 EFORT

15.2
The Problem: Wear Debris Particles

Although the etiology of aseptic loosening is multifactorial and complex, one of the generally accepted contributions to this time-dependent process arises from cumulative and interrelated cell-mediated reactions to wear debris particles [3, 4, 6, 7]. Originally this process was termed "cement disease," as it was thought to be a response to particles of polymethylmethacrylate (PMMA) bone cement, but later it was observed that the process also occurs with cementless implants [1]. The high level of activity and the satisfactory quality of life of patients with hip prostheses accounts for a higher wear rate and consequently an increase in the quantity of wear particles produced [1, 8]. General data indicate that wear rate is related to the survival of the implant and that the response to debris depends on particles and possibly on patient-related factors such as material sensitivity or genetic predisposition [3, 7, 9].

The type of prosthesis, method of fixation and bearing surfaces used are known factors that have a significant impact on the development of osteolysis [2, 10]. The most common materials used in THR consist of an ultra-high-molecular-weight polyethylene (UHMWPE) cup articulating upon a metal or ceramic femoral head [2]. Different bearing surface combinations produce different quantities of wear particles, e.g., polyethylene wear on metal is reported to be 5–10 times higher than polyethylene versus ceramic [11, 12]. The problems associated with the degradation of the polyethylene component have led to the introduction of new materials that effectively reduce the total amount of wear generated, such as combinations of metal-on-metal and ceramic-on-ceramic materials and, most recently, cross-linked polyethylenes for articulating bearing surfaces [1, 2, 13–15]. In addition to polyethylene wear particles, metallic and ceramic particles released may initiate a cellular response that contributes to periprosthetic bone resorption [3, 4]. Moreover, wear particles might cause third-body wear, which accelerates the degradation of the prosthetic material [3, 16].

UHMWPE particles appear in periprosthetic tissues in close association with activated macrophages and giant cells. Studies of UHMWPE particles recovered from retrieved membranes have reported that their shape is quite irregular and with a micrometric size [17]. The amount of metallic particles found in these tissues is greater than the number of UHMWPE particles, though their size is much smaller and the shape more regular [18]. The smaller dimensions of metallic debris, which may fall in the nanometric size range, facilitate not only activation of cells in the periprosthetic space but also the access of particles to distant organs, and the release of ions due to the increased total surface area, facilitating corrosion in the presence of aggressive biological fluids [7, 17]. Osteolysis related to the presence of ceramic debris has rarely been reported, although some authors have observed that in cases of abnormally high wear rates, particles may initiate a foreign body inflammatory response [19]. Analysis from retrieved tissues has revealed the spheroid-shaped and submicrometric-sized particles of ceramics, overlapping the size range commonly observed in metal and UHMWPE [19].

The access of debris to the effective joint space and the pressure of the joint fluid ensure the contact of particles with cells located in the periprosthetic space, which will become activated. The resulting bone erosion eases the entrance of joint fluid and particles into the interface, gradually increasing and leading to the progressive loosening of the implant [3]. In the femoral component, improvements in coating techniques have reduced the incidence of osteolysis by blocking access of wear particles [20]. However, in the acetabular area, wear debris can reach the interface through unfilled screw holes or via non-ingrown areas of the shell [21].

Both resident and recruited cells are actively involved in the progression of wear particle-induced osteolysis, such as macrophages and giant cells, osteoblasts, osteoclasts, fibroblasts and lymphocytes [6, 9, 22]. As a result of the study of biological mechanisms implicated in this pathology, novel cells and molecules are emerging as potential mediators, illustrating the complexity of this process. Cellular activation triggers the expression and secretion of a vast array of potent mediators including chemokines, growth factors, pro- and anti-inflammatory cytokines, degradative enzymes, reactive oxygen radicals and other molecules which recruit and stimulate cells capable of inducing bone resorption and fibrous tissue formation [17, 23]. As result of this chronic inflammatory response, a pseudomembrane at the bone-implant interface is formed, which compromises the osseointegration of the device, undermining the periprosthetic bed and favoring the access of particles to the effective joint space, self-perpetuating the inflammatory response (Fig.15.1).

Generally, the cellular effects due to the presence of particles can be classified into direct and indirect effects, with the latter mediated by soluble factors released by cells exposed to wear particles.

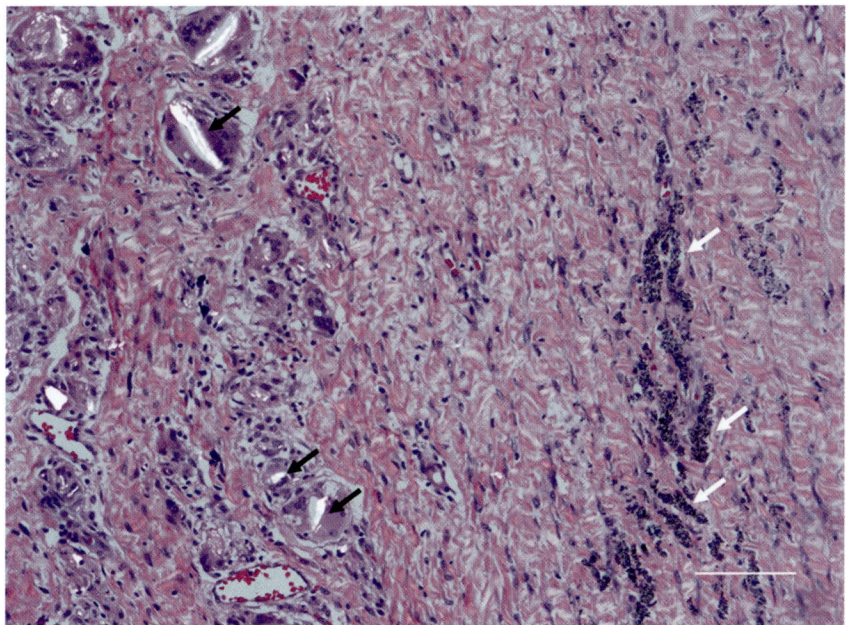

Fig. 15.1 Periprosthetic membrane retrieved from a patient undergoing revision surgery to replace a loosened hip prosthesis. Microscopy imaging showing the characteristic features of the foreign body reaction induced by wear debris particles, with an infiltration of inflammatory cells. UHWMPE particles, larger, corresponding to bright areas, are found associated with multinuclear giant cells whereas smaller metallic particles, corresponding to dark areas are found in macrophages. *Black* and *white arrows* indicate the presence of UHWMPE and metallic particles, respectively. Hematoxylin-Eosin staining. Bar: 100 μm

15.3
Direct Effects of Wear Debris Particles

The involvement of macrophages as primary target cells for particles is widely known, as they are capable of systemic migration to sites containing debris [24, 25]. A recent study reported that murine monocytes and bone marrow-derived macrophages exposure to low amounts of metallic particles may contribute to cell survival, which may account for early events preceding the osteolytic process [26]. Ren et al. have reported that the transient depletion of macrophages prevent UHMWPE particle-induced inflammation, supporting the main role of these cells in this process [27]. The participation of macrophages is not only related to the initial response events, since they are able to amplify the inflammatory response to circulating monocytes and inflammatory cells [27, 28]. Moreover, macrophages have generated interest recently, as they share a common precursor lineage with osteo-clasts. *In vitro*, several extracellular signals, including the presence of soluble factors and the phagocytosis of particles, induce differentiation from macrophages to bone-resorbing osteoclasts [29]. The main action attributed to macrophages in the osteolytic response has been associated with the ingestion of particles and the related subsequent effects, including cell damage and release of pro-resorptive factors, including tumor necrosis factor (TNF-α), interleukin-1β (IL-1β), interleukin-6 (IL-6), prostaglandin E$_2$ (PGE$_2$) and granulocyte-macrophage colony-stimulating factor (GM-CSF) [22, 24, 25]. These cytokines not only promote osteoclastogenesis but also are able to interfere with osteogenesis guided by osteoprogenitor cells. Lysis of macrophages releases ingested particles into the extracel-lular milieu that can be taken up by other macrophages, and also cell components that may stimulate other cells [30]. Clinical and animal studies have related the progression of oste-olysis with the continuous supply of wear debris that arises from the prosthesis, which would counterbalance a possible "inactivation" of particles after phagocytosis by mac-rophages [31]. Recently, it has been established that metallic wear particles exert cytotoxic effects in macrophages, including DNA damage, oxidative stress and release of soluble mediators [7]. If the particle is too large to be internalized, as it is the case of polyethylene particles, then macrophages can fuse in an attempt to isolate the particle, forming large, multinucleated foreign body giant cells [22, 28].

Several reports have indicated that activation and release of pro-inflammatory cytokines is dependent on a membrane-particle recognition process, independently of phagocytosis [6, 32]. In this regard, some authors have focused on the molecular mechanisms involved in particle recognition such as opsonization and cell surface receptors [33, 34]. Recently it has been reported that Toll-like receptor (TLR) signaling pathway acts by mediating particle-activated macrophages [30]. Other authors have suggested that activation through TLR is dependent on the presence of bacterial components or endogenous ligands released by necrotic or activated cells adhered to wear particles, even in the absence of any clinical signs of infection [35]. The relative contribution of lipopolysaccharide to aseptic loosening still is uncertain. Several authors have performed *in vitro* studies to analyze the affinity of endotoxin binding to different particles and the consequences on macrophage activation [36, 37]. CD14 is a specific cell membrane receptor of macrophages and cofactor for TLR-mediated signaling, and it has been proposed to be a receptor for the stress-inducible 72-kDa

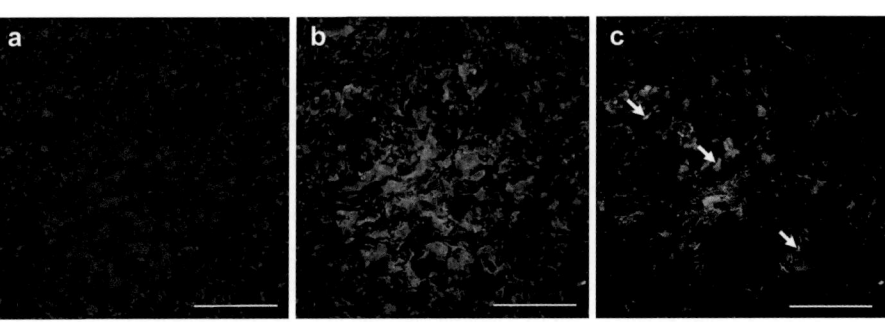

Fig. 15.2 Expression of CD14 and Hsp72 in a periprosthetic membrane retrieved from a patient undergoing revision surgery for aseptic loosening of the Ti-based cup. Confocal single projections showing CD14 (**a**, red), Hsp72 (**b**, green) and overlaid images of CD14 and Hsp72 (**c**, yellow). Arrows in **c** show co-localization of CD14 and Hsp72. DAPI staining was used for counterstaining purposes. Bar: 150 mm

form of the heat shock protein 70, Hsp72. Studies from Asea et al. have reported that extracellular Hsp72 acts as a cytokine, triggering the secretion of TNF-α and IL-1β through a CD14-mediated mechanism [38]. More recently, we have found that CD14 and Hsp72 co-localize in periprosthetic membranes retrieved from patients undergoing hip revision surgery due to aseptic loosening of titanium (Ti)-based cups (Fig. 15.2). Employing an *in vitro* model, we observed that extracellular Hsp72 amplifies the inflammatory response to wear debris by further increasing the secretion levels of TNF-α and IL-1β induced by Ti-particles [39]. Apart from providing additional evidence that novel target molecules and their interactions deserve further analysis, these data suggest that the etiology of aseptic loosening involves signaling pathways that are not yet well understood.

As mentioned, size is a key feature in establishing the potential bioreactivity of wear particles. To elicit an *in vitro* cellular response, particles need to be in the phagocytosable range (<10 μm-sized particles). Although a threshold bioreactivity value has not been defined, particles with smaller sizes are likely to induce a stronger inflammatory response as compared to the larger ones [7, 40]. In addition to size, cellular toxicity and intensity of the inflammatory response also depend on several intrinsic and interrelated particle features such as shape, surface, topography and chemistry, which influence the interaction of particles with the biological environment [7, 24]. The release of pro-inflammatory cytokines from macrophages after exposure to Ti particles has been reported by many authors [25, 40]. Upon stimulation with Ti particles, murine bone marrow macrophages differentiation into osteoclasts is enhanced in the presence of macrophage colony-stimulating factor (M-CSF) and receptor activator for nuclear factor κB ligand (RANKL) [41]. The advances at the cellular level have focused attention on molecular mechanisms involved in this response, such as the activation of mitogen-activated protein kinases (MAPK), transcription factors as nuclear factor κB (NF-kB) or cytokines signaling, including the implication of inflammasome [7, 42]. Comparisons between different kinds of particles is an important issue, thus we were prompted to comparatively study the inflammatory response to metallic and ceramic Ti-based particles, which would arise as degradation of modified surface implants [40]. Results obtained using *in vitro* macrophage cell culture models showed the relative inertness of the ceramic, rutile (TiO_2) particles, versus Ti particles.

Particles may also affect correct bone tissue remodeling around the implant, increasing the recruitment, differentiation and activity of osteoclasts (bone-resorptive multinucleated cells derived from hematopoietic stem cells) and/or disturbing the differentiation, functionality and activity of osteoblasts (bone-forming cells derived from bone marrow stromal cells), both of which flank the bone-implant interface [43]. The long-term success of a prosthetic device is based on its correct osseointegration and the long-term quality of periprosthetic bone tissues is dependent on the recruitment, attachment and differentiation of osteogenic cells. The involvement of osteoblastic cells in osteolysis is based on their bone-forming ability and their direct role in the osteoclastogenesis process through the axis OPG-RANK-RANKL [6, 17]. These molecules are susceptible to modulation by several factors, including exposure to wear particles and soluble mediators released by macrophages activated by the presence of wear particles, which may also synergize their effects. The effects of wear particles have been observed on adhesion, proliferation, cytotoxicity, prostanoid production (PGE_2), chemokines (IL-8 and monocyte chemotactic protein-1 (MCP-1)), pro-resorptive factors (IL-6 and RANKL) and pro-osteogenic (transforming growth factor-β (TGF-β)), negative regulators of differentiation (bone morphogenetic protein-3 and sclerostin) and osteoblasts markers such as alkaline-phosphatase-specific activity, osteonectin and osteocalcin [23, 44–47]. Their biological effects are dependent on the size, concentration, shape and composition of the particles [48, 49]. Indeed, many of the osteoblastic responses may be mediated by the internalization of particles in the phagocytosable range, which may account for the expression of specific surface receptors, as CD68 [50]. *In vitro* studies from our group have shown that the internalization of Ti particles may impair the osteoblast-substrate attachment, through alterations in the cytoskeleton, release of osteoblastic mediators including IL-6, PGE_2 and GM-CSF and down-regulation of the secretion of osteoprotegerin (OPG) [40, 46]. Other authors have reported that Ti particles are also able to reduce procollagen synthesis by osteoblastic cells, inhibiting the bone matrix formation [51]. The specific molecular effects triggered by phagocytosable Ti particles involve the activation of multiple signaling pathways such as protein tyrosine kinases (PTKs) and protein kinase C (PKC) pathways, resulting in altered gene expression via the activation of nuclear transcription factors including NFkB and nuclear factor IL-6 (NFIL-6) [43, 51].

The prolonged accumulation of wear-debris particles also facilitates their contact with osteoprogenitor cells, compromising their differentiation and limiting the reservoir of bone-forming cells. *In vitro* effects of particles have been evaluated on proliferation, viability, and differentiation potential of mesenchymal stem cells [43, 52]. Interestingly, these effects could be also mediated via particle internalization [43, 53].

The high intracapsular pressure observed in patients with loosened implants have focused attention on cells responsive to changes in pressure, including osteocytes, osteoblasts, macrophages and osteoclastic precursors [54, 55]. In the bone microenvironment, osteocytes, which are terminally differentiated cells of the osteoblastic lineage and regulate bone remodeling, may also be responsive to wear particles and consequently participate in the periprosthetic bone resorption [56, 57]. Recent research has reported the involvement of osteocytes in particle-induced inflammation and bone erosion by inducing the release of cytokines [56].

15.4
Indirect Effects of Wear Debris Particles: Mediation by Soluble Factors

It is generally accepted that osteolysis is unlikely to be caused by only one particular cell type, but rather by the conjunction of several cellular reactions to wear particles and inter-related factors that collectively contribute to periprosthetic bone loss. Due to the chemot-actic properties of chemokines, including MCP-1 and macrophage inflammatory protein 1-α (MIP-1α), systemic monocytes and osteoclastic precursors are attracted to the site of inflammation [58]. The effects mediated by soluble factors released by osteoblasts and macrophages exposed to particles on osteoclastogenesis have been well documented *in vitro* [59]. Conditioned media from monocytes-macrophages exposed to particles is able to induce bone resorption in a model of *ex vivo* cultured rat calvaria [60].

In the periprosthetic milieu, macrophages and osteoblasts release soluble factors able to act in paracrine and autocrine fashions to modulate the response initiated by wear particles. Many *in vitro* studies have shown the interrelations and modulations between factors involved in particle-induced osteolysis [61–64]. These studies reveal that the interrelation between cells mediated by soluble factors should be taken into account for effective extrap-olation to the *in vivo* situation. Earlier experiments that employed conditioned media or recombinant proteins provided scarce information about the sustained and bidirectional interaction between both kinds of cells [65, 66]. The later establishment of co-culture sys-tems overcame these deficiencies, providing useful information about the inflammatory response induced by particles of different composition [63, 65, 67]. Data derived from our group support the idea that osteoblasts are able to modulate the inflammatory response initi-ated by activated macrophages [61]. Co-culture experiments have also provided additional information about IL-1β as one of the main mediators involved in the amplification of the inflammatory response elicited by wear debris particles in macrophages [61, 68]. Although this *in vitro* information should be cautiously interpreted, since in the periprosthetic space there are wear debris particles of very different nature, many coexisting cells and molecules susceptible to feedback loops, it provides a valuable tool for further studies.

References

1. Learmonth, I.D., Young, C., Rorabeck, C.: The operation of the century: total hip replacement. Lancet **370**, 1508–1519 (2007). doi:10.1016/S0140-6736(07)60457-7
2. Passuti, N., Philippeau, J.M., Gouin, F.: Friction couples in total hip replacement. Orthop. Traumatol. Surg. Res. **95**, S27–S34 (2009). doi:10.1016/j.otsr.2009.04.003
3. Sundfeldt, M., Carlsson, L.V., Johansson, C.B., Thomsen, P., Gretzer, C.: Aseptic loosening, not only a question of wear: a review of different theories. Acta Orthop. **77**, 177–197 (2006). doi:10.1080/17453670610045902
4. Catelas, I., Jacobs, J.J.: Biologic activity of wear particles. Instr. Course Lect. **59**, 3–16 (2010)
5. Lübbeke, A., Katz, J.N., Perneger, T.V., Hoffmeyer, P.: Primary and revision hip arthroplasty: 5-year outcomes and influence of age and comorbidity. J. Rheumatol. **34**, 394–400 (2007)

6. Abu-Amer, Y., Darwech, I., Clohisy, J.C.: Aseptic loosening of total joint replacements: mechanisms underlying osteolysis and potential therapies. Arthritis Res. Ther. **9**(Suppl 1), S6 (2007). doi:doi: 10.1186/ar2170

7. Hallab, N.J., Jacobs, J.J.: Biologic effects of implant debris. Bull NYU Hosp. Jt. Dis. **67**, 182–188 (2009)

8. Schmalzried, T.P., Huk, O.L.: Patient factors and wear in total hip arthroplasty. Clin. Orthop. Relat. Res. **418**, 94–97 (2004)

9. Tuan, R.S., Lee, F.Y.I., Konttinen, Y.T., Wilkinson, J.M., Smith, R.L.: What are the local and systemic biologic reactions and mediators to wear debris and what host factors determine or modulate the biologic response to wear particles? J. Am. Acad. Orthop. Surg. **16**, S42–S48 (2008)

10. Purdue, P.E., Koulouvaris, P., Nestor, B.J., Sculco, T.P.: The central role of wear debris in periprosthetic osteolysis. HSS J. **2**, 102–113 (2006). doi:10.1007/s11420-006-9003-6

11. Skinner, H.B.: Ceramic bearing surfaces. Clin. Orthop. Relat. Res. **369**, 83–91 (1999)

12. Affatato, S., Spinelli, M., Zavalloni, M., Traina, F., Carmignato, S., Toni, A.: Ceramic-on-metal for total hip replacement: mixing and matching can lead to high wear. Artif. Organs **34**, 319–323 (2010). doi:10.1111/j.1525-1594.2009.00854.x

13. D'Antonio, J.A., Sutton, K.: Ceramic materials as bearing surfaces for total hip arthroplasty. J. Am. Acad. Orthop. Surg. **17**, 63–68 (2009)

14. García-Cimbrelo, E., García-Rey, E., Murcia-Mazón, A., Blanco-Pozo, A., Martí, E.: Alumina-on-alumina in THA: a multicenter prospective study. Clin. Orthop. Relat. Res. **466**, 309–316 (2008). doi:10.1007/s11999-007-0042-1

15. García-Rey, E., García-Cimbrelo, E., Cruz-Pardos, A., Ortega-Chamarro, J.: New polyethylenes in total hip replacement. A prospective comparative clinical study of two types of liner. J. Bone Joint Surg. Br. **90**, 149–153 (2008). doi:DOI: 10.1302/0301-620X.90B2.19887

16. Brown, T.D., Lundberg, H.J., Pedersen, D.R., Callaghan, J.J.: 2009 Nicolas Andry award: clinical biomechanics of third body acceleration of total hip wear. Clin. Orthop. Relat. Res. **467**, 1885–1897 (2009). doi:10.1007/s11999-009-0854-2

17. Holt, G., Murnaghan, C., Reilly, J., Meek, R.M.: The biology of aseptic osteolysis. Clin. Orthop. Relat. Res. **460**, 240–252 (2007). doi:10.1097/BLO.0b013e31804b4147

18. Doorn, P.F., Campbell, P.A., Worrall, J., Benya, P.D., McKellop, H.A., Amstutz, H.C.: Metal wear particle characterization from metal on metal total hip replacements: transmission electron microscopy study of periprosthetic tissues and isolated particles. J. Biomed. Mater. Res. **42**, 103–111 (1998)

19. Hatton, A., Nevelos, J.E., Nevelos, A.A., Banks, R.E., Fisher, J., Ingham, E.: Alumina-alumina artificial hip joints. Part I: a histological analysis and characterisation of wear debris by laser capture microdissection of tissues retrieved at revision. Biomaterials **23**, 3429–3440 (2002)

20. Dattani, R.: Femoral osteolysis following total hip replacement. Postgrad. Med. J. **83**, 312–316 (2007). doi:10.1136/pgmj.2006.053215

21. Agarwal, S.: Osteolysis-basic science, incidence and diagnosis. Curr. Orthop. **18**, 220–231 (2004). doi:10.1016/j.cuor.2004.03.002

22. Gallo, J., Raska, M., Mrázek, F., Petrek, M.: Bone remodeling, particle disease and individual susceptibility to periprosthetic osteolysis. Physiol. Res. **57**, 339–349 (2008)

23. Goodman, S.B., Ma, T.: Cellular chemotaxis induced by wear particles from joint replacements. Biomaterials **31**, 5045–5050 (2010). doi:10.1016/j.biomaterials.2010.03.046

24. Drees, P., Eckardt, A., Gay, R.E., Gay, S., Huber, L.C.: Mechanisms of disease: molecular insights into aseptic loosening of orthopedic implants. Nat. Clin. Pract. Rheumatol. **3**, 165–171 (2007). doi:10.1038/ncprheum0428

25. Ingham, E., Fisher, J.: The role of macrophages in osteolysis of total joint replacement. Biomaterials **26**, 1271–1286 (2005). doi:10.1016/j.biomaterials.2004.04.035

26. Lacey, D.C., De Kok, B., Clanchy, F.I., Bailey, M.J., Speed, K., Haynes, D., Graves, S.E., Hamilton, J.A.: Low dose metal particles can induce monocyte/macrophage survival. J. Orthop. Res. **27**, 1481–1486 (2009). doi:10.1002/jor.20914

27. Ren, W., Markel, D.C., Schwendener, R., Ding, Y., Wu, B., Wooley, P.H.: Macrophage depletion diminishes implant-wear-induced inflammatory osteolysis in a mouse model. J. Biomed. Mater. Res. A **85**, 1043–1051 (2008). doi:10.1002/jbm.a.31665

28. Revell, P.A.: The combined role of wear particles, macrophages and lymphocytes in the loosening of total joint prostheses. J. R. Soc. Interface **5**, 1263–1278 (2008). doi:10.1098/rsif.2008.0142

29. Fujikawa, Y., Itonaga, I., Kudo, O., Hirayama, T., Taira, H.: Macrophages that have phagocytosed particles are capable of differentiating into functional osteoclasts. Mod. Rheumatol. **15**, 346–351 (2005). doi:10.1007/s10165-005-0424-8

30. Maitra, R., Clement, C.C., Scharf, B., Crisi, G.M., Chitta, S., Paget, D., Purdue, P.E., Cobelli, N., Santambrogio, L.: Endosomal damage and TLR2 mediated inflammasome activation by alkane particles in the generation of aseptic osteolysis. Mol. Immunol. **47**, 175–184 (2009). doi:10.1016/j.molimm.2009.09.023

31. Xing, Z., Schwab, L.P., Alley, C.F., Hasty, K.A., Smith, R.A.: Titanium particles that have undergone phagocytosis by macrophages lose the ability to activate other macrophages. J. Biomed. Mater. Res. B Appl. Biomater. **85**, 37–41 (2008). doi:10.1002/jbm.b.30913

32. Nakashima, Y., Sun, D.H., Trindade, M.C., Maloney, W.J., Goodman, S.B., Schurman, D.J., Smith, R.L.: Signaling pathways for tumor necrosis factor-alpha and interleukin-6 expression in human macrophages exposed to titanium-alloy particulate debris *in vitro*. J. Bone Joint Surg. Am. **81**, 603–615 (1999)

33. Rakshit, D.S., Lim, J.T., Ly, K., Ivashkiv, L.B., Nestor, B.J., Sculco, T.P., Purdue, P.E.: Involvement of complement receptor 3 (CR3) and scavenger receptor in macrophage responses to wear debris. J. Orthop. Res. **24**, 2036–2044 (2006). doi:10.1002/jor.20275

34. Zolotarevová, E., Hudeček, J., Spundová, M., Entlicher, G.: Binding of proteins to ultra high molecular weight polyethylene wear particles as a possible mechanism of macrophage and lymphocyte activation. J. Biomed. Mater. Res. A **95**(3), 950–955 (2010)

35. Lähdeoja, T., Pajarinen, J., Kouri, V.P., Sillat, T., Salo, J., Konttinen, Y.T.: Toll-like receptors and aseptic loosening of hip endoprosthesis-a potential to respond against danger signals? J. Orthop. Res. **28**, 184–190 (2010). doi:10.1002/jor.20979

36. Smith, R.A., Hallab, N.J.: *In vitro* macrophage response to polyethylene and polycarbonate-urethane particles. J. Biomed. Mater. Res. A **93**, 347–355 (2010). doi:10.1002/jbm.a.32529

37. Wilkins, R., Tucci, M., Benghuzzi, H.: Evaluation of endotoxin binding to uhmwpe and inflammatory mediator production by macrophages. Biomed. Sci. Instrum. **44**, 459–464 (2008)

38. Asea, A., Kraeft, S.K., Kurt-Jones, E.A., Stevenson, M.A., Chen, L.B., Finberg, R.W., Koo, G.C., Calderwood, S.K.: HSP70 stimulates cytokine production through a CD14-dependant pathway, demonstrating its dual role as a chaperone and cytokine. Nat. Med. **6**, 435–442 (2000)

39. Vallés, G., Vilaboa, N., Munuera, L., García-Cimbrelo, E.: Hsp72: a new mediator in wear particles-induced osteolysis. 11th European Federation of National Associations of Orthopaedics and Traumatology Congress (EFFORT), 2–5 June, Madrid (2010)

40. Vallés, G., González-Melendi, P., González-Carrasco, J.L., Saldaña, L., Sánchez-Sabaté, E., Munuera, L., Vilaboa, N.: Differential inflammatory macrophage response to rutile and titanium particles. Biomaterials **27**, 5199–5211 (2006). doi:10.1016/j.biomaterials.2006.05.045

41. Liu, F., Zhu, Z., Mao, Y., Liu, M., Tang, T., Qiu, S.: Inhibition of titanium particle-induced osteoclastogenesis through inactivation of NFATc1 by VIVIT peptide. Biomaterials **30**, 1756–1762 (2009). doi:10.1016/j.biomaterials.2008.12.018

42. Beidelschies, M.A., Huang, H., McMullen, M.R., Smith, M.V., Islam, A.S., Goldberg, V.M., Chen, X., Nagy, L.E., Greenfield, E.M.: Stimulation of macrophage TNFalpha production by

orthopaedic wear particles requires activation of the ERK1/2/Egr-1 and NF-kappaB pathways but is independent of p38 and JNK. J. Cell. Physiol. **217**, 652–666 (2008). doi:10.1002/jcp. 21539

43. Goodman, S.B., Ma, T., Chiu, R., Ramachandran, R., Smith, R.L.: Effects of orthopaedic wear particles on osteoprogenitor cells. Biomaterials **27**, 6096–6101 (2006). doi:10.1016/j. biomaterials.2006.08.023

44. Vallés, G., González-Melendi, P., Saldaña, L., Rodriguez, M., Munuera, L., Vilaboa, N.: Rutile and titanium particles differentially affect the production of osteoblastic local factors. J. Biomed. Mater. Res. A **84**, 324–336 (2008). doi:10.1002/jbm.a.31315

45. Ma, G.K., Chiu, R., Huang, Z., Pearl, J., Ma, T., Smith, R.L., Goodman, S.B.: Polymethylmethacrylate particle exposure causes changes in p38 MAPK and TGF-beta signaling in differentiating MC3T3-E1 cells. J. Biomed. Mater. Res. A **94**, 234–240 (2010). doi:10.1002/jbm.a.32686

46. Saldaña, L., Vilaboa, N.: Effects of micrometric titanium particles on osteoblast attachment and cytoskeleton architecture. Acta Biomater. **6**, 1649–1660 (2010). doi:10.1016/j.actbio. 2009.10.033

47. Granchi, D., Amato, I., Battistelli, L., Ciapetti, G., Pagani, S., Avnet, S., Baldini, N., Giunti, A.: Molecular basis of osteoclastogenesis induced by osteoblasts exposed to wear particles. Biomaterials **26**, 2371–2379 (2005). doi:10.1016/j.biomaterials.2004.07.0145

48. Lohmann, C.H., Dean, D.D., Köster, G., Casasola, D., Buchhorn, G.H., Fink, U., Schwartz, Z., Boyan, B.D.: Ceramic and PMMA particles differentially affect osteoblast phenotype. Biomaterials **23**, 1855–1863 (2002)

49. Choi, M.G., Koh, H.S., Kluess, D., O'Connor, D., Mathur, A., Truskey, G.A., Rubin, J., Zhou, D.X., Sung, K.L.: Effects of titanium particle size on osteoblast functions *in vitro* and *in vivo*. Proc. Natl Acad. Sci. USA **102**, 4578–4583 (2005). doi:10.1073/pnas.0500693102

50. Heinemann, D.E., Lohmann, C., Siggelkow, H., Alves, F., Engel, I., Köster, G.: Human osteoblast-like cells phagocytose metal particles and express the macrophage marker CD68 *in vitro*. J. Bone Joint Surg. Br. **82**, 283–289 (2000)

51. Vermes, C., Roebuck, K.A., Chandrasekaran, R., Dobai, J.G., Jacobs, J.J., Glant, T.T.: Particulate wear debris activates protein tyrosine kinases and nuclear factor kappaB, which down-regulates type I collagen synthesis in human osteoblasts. J. Bone Miner. Res. **15**, 1756–1765 (2000)

52. Chiu, R., Ma, T., Smith, R.L., Goodman, S.B.: Ultrahigh molecular weight polyethylene wear debris inhibits osteoprogenitor proliferation and differentiation *in vitro*. J. Biomed. Mater. Res. A **89**, 242–247 (2009). doi:10.1002/jbm.a.32001

53. Schofer, M.D., Fuchs-Winkelmann, S., Kessler-Thönes, A., Rudisile, M.M., Wack, C., Paletta, J.R., Boudriot, U.: The role of mesenchymal stem cells in the pathogenesis of Co-Cr-Mo particle induced aseptic loosening: an *in vitro* study. Biomed. Mater. Eng. **18**, 395–403 (2008). doi:10.3233/BME-2008-0556

54. McEvoy, A., Jeyam, M., Ferrier, G., Evans, C.E., Andrew, J.G.: Synergistic effect of particles and cyclic pressure on cytokine production in human monocyte/macrophages: proposed role in periprosthetic osteolysis. Bone **30**, 171–177 (2002)

55. Tan, S.D., de Vries, T.J., Kuijpers-Jagtman, A.M., Semeins, C.M., Everts, V., Klein-Nulend, J.: Osteocytes subjected to fluid flow inhibit osteoclast formation and bone resorption. Bone **41**, 745–751 (2007). doi:10.1016/j.bone.2007.07.019

56. Kanaji, A., Caicedo, M.S., Virdi, A.S., Sumner, D.R., Hallab, N.J., Sena, K.: Co-Cr-Mo alloy particles induce tumor necrosis factor alpha production in MLO-Y4 osteocytes: a role for osteocytes in particle-induced inflammation. Bone **45**, 528–533 (2009). doi:10.1016/j. bone.2009.05.020

57. Atkins, G.J., Welldon, K.J., Holding, C.A., Haynes, D.R., Howie, D.W., Findlay, D.M.: The induction of a catabolic phenotype in human primary osteoblasts and osteocytes by polyethylene particles. Biomaterials **30**, 3672–3681 (2009). doi:10.1016/j.biomaterials.2009.03.035

58. Huang, Z., Ma, T., Ren, P.G., Smith, R.L., Goodman, S.B.: Effects of orthopedic polymer particles on chemotaxis of macrophages and mesenchymal stem cells. J. Biomed. Mater. Res. A **94**, 1264–1269 (2010). doi:10.1002/jbm.a.32803

59. Greenfield, E.M., Bi, Y., Ragab, A.A., Goldberg, V.M., Van De Motter, R.R.: The role of osteoclast differentiation in aseptic loosening. J. Orthop. Res. **20**, 1–8 (2002)

60. Ren, W., Wu, B., Mayton, L., Wooley, P.H.: Polyethylene and methyl methacrylate particle-stimulated inflammatory tissue and macrophages up-regulate bone resorption in a murine neonatal calvaria *in vitro* organ system. J. Orthop. Res. **20**, 1031–1037 (2002)

61. Vallés, G., Gil-Garay, E., Munuera, L., Vilaboa, N.: Modulation of the cross-talk between macrophages and osteoblasts by titanium-based particles. Biomaterials **29**, 2326–2335 (2008). doi:10.1016/j.biomaterials.2008.02.011

62. Park, Y.G., Kang, S.K., Kim, W.J., Lee, Y.C., Kim, C.H.: Effects of TGF-beta, TNF-alpha, IL-beta and IL-6 alone or in combination, and tyrosine kinase inhibitor on cyclooxygenase expression, prostaglandin E2 production and bone resorption in mouse calvarial bone cells. Int. J. Biochem. Cell Biol. **36**, 2270–2280 (2004). doi:10.1016/j.biocel.2004.04.019

63. Horowitz, S.M., Gonzales, J.B.: Inflammatory response to implant particulates in a macrophage/osteoblast coculture model. Calcif. Tissue Int. **59**, 392–396 (1996)

64. Liu, X.H., Kirschenbaum, A., Yao, S., Levine, A.C.: Cross-talk between the interleukin-6 and prostaglandin E(2) signaling systems results in enhancement of osteoclastogenesis through effects on the osteoprotegerin/receptor activator of nuclear factor-{kappa}B (RANK) ligand/RANK system. Endocrinology **146**, 1991–1998 (2005). doi:10.1210/en.2004-1167

65. Zreiqat, H., Crotti, T.N., Howlett, C.R., Capone, M., Markovic, B., Haynes, D.R.: Prosthetic particles modify the expression of bone-related proteins by human osteoblastic cells *in vitro*. Biomaterials **24**, 337–346 (2003)

66. Horowitz, S.M., Rapuano, B.P., Lane, J.M., Burstein, A.H.: The interaction of the macrophage and the osteoblast in the pathophysiology of aseptic loosening of joint replacements. Calcif. Tissue Int. **54**, 320–324 (1994)

67. Rodrigo, A., Vallés, G., Saldaña, L., Rodríguez, M., Martínez, M.E., Munuera, L., Vilaboa, N.: Alumina particles influence the interactions of cocultured osteoblasts and macrophages. J. Orthop. Res. **24**, 46–54 (2006). doi:10.1002/jor.20007

68. St Pierre, C.A., Chan, M., Iwakura, Y., Ayers, D.C., Kurt-Jones, E.A., Finberg, R.W.: Periprosthetic osteolysis: characterizing the innate immune response to titanium wear-particles. J. Orthop. Res. **28**, 1418–1424 (2010). doi:10.1002/jor.21149

Part V

Miscellaneous

Cushion Form Bearings in Total Hip Arthroplasty: Nature's Approach to the Synovial Joint Problem

16

Antonio Moroni, Martha Hoque, Giovanni Micera, Riccardo Orsini, Emanuele Nocco, and Sandro Giannini

16.1
Introduction

Total hip replacement (THR) is an effective surgical procedure for patients with a variety of hip diseases. Although patient satisfaction is high, implant longevity remains a cause for concern [1]. The most common source of implant failure is mechanical loosening [2], which may be due to inadequate primary fixation and/or loss of fixation at medium–long term. The latter is generally caused by peri-implant osteolysis which is induced by wear debris produced during movement [3, 4]. There is also growing clinical evidence that the total volume of wear debris has a great effect on loosening [5, 6].

Over the last decades, many efforts have been successfully made to improve implant fixation and design, but the search for the optimal bearing type is ongoing. Although there is no consensus on the best type of implant, selection of the bearing surface is the crucial factor for a successful, durable THR, particularly in young, active individuals. The most commonly used bearing surface is metal-on-ultra-high-molecular-weight polyethylene (UHMWPE). However, with UHMWPE, the generation of wear debris which leads to mechanical loosening and consequently revision surgery, is a significant drawback [7]. Cross-linked UHMWPE has recently been developed in an effort to improve the wear characteristics of the standard material. Although in vitro studies have confirmed decreased wear rates with cross-linked UHMWPE compared to standard UHMWPE, there have nonetheless been reports of mechanical failures including rim cracking and delaminations [8, 9].

A. Moroni (✉), M. Hoque, G. Micera, R. Orsini, and S. Giannini
Department of Orthopaedics and Traumatology, Bologna University,
Via G. C. Pupilli 1, 40136 Bologna, Italy
e-mail: a.moroni@ior.it

E. Nocco
Clinical Research and Development Active Implants Corporation,
5865 Ridgeway Center Parkway, Suite 218,
Memphis, TN 38120, USA

K. Knahr (ed.), *Tribology in Total Hip Arthroplasty*,
DOI: 10.1007/978-3-642-19429-0_16, © 2011 EFORT

Wear is extremely low with ceramic-on-ceramic (COC) bearings. However, even this bearing type has drawbacks, including head and liner ruptures, noisy hips and foreign body reactions caused by ceramic wear [10].

It is well known that large diameter prosthetic heads allow for an increased range of motion and stability, but both metal-on-UHMWPE and COC bearings comprise prosthetic heads which are significantly smaller than the patient's original head size. This is due to technological limitations associated with these bearing types.

The original size of the patient's femoral head can be replicated by using metal-on-metal (MOM) bearings. Another advantage of this bearing type is low wear [11]. However, there are also drawbacks associated with MOM bearings, including postoperative metal ion elevation in the serum, the development of hypersensitivity, the formation of pseudo-tumors, and increased wear in cases of malpositioning [12].

Therefore, in hip arthroplasty, there is a need for new bearing technology which could better replicate the anatomical and functional characteristics of the normal hip joint.

16.2
Cushion Form Bearings

The natural synovial joint provides low wear because of the maintenance of a fluid film that considerably reduces friction by separating the articulating surfaces [13]. The development of a fluid film is predominantly due to a combination of elasto-hydrodynamic lubrication (EHL) and micro-elasto-hydrodynamic lubrication (μEHL) [14]. Elasto-hydrodynamic lubrication is produced when the pressure developed in a converging film of lubricant between two articulating surfaces (e.g. cartilage) is sufficient to cause local elastic deformation of either surface, keeping them separated [15]. Micro-elasto-hydrodynamic lubrication is a localized form of EHL, whereby pressure perturbations cause substantial flattening of asperities at the material's surfaces, increasing conformity and assisting in the maintenance of a lubricious film [16].

Unsworth et al. [17] showed that if an artificial joint was constructed like a human joint, then it too could benefit from fluid-film lubrication just as human joints do. The theory behind this was conceived in 1986, when Dowson and Jin [18] developed their micro-elasto-hydrodynamic lubrication theory, and Unsworth et al. [19] studied factors which led to this very low friction. Using the hip friction simulator and compliant layered joints, Burgess [20] plotted the friction factors in a Stribeck curve and compared these experimentally to obtain values with both the calculated EHL and μEHL values [21].

Therefore, it seems that cushion form bearings operate with negligible asperity contact and wear should thus be reduced. This was borne out in work by Smith et al. [22] who found that the average wear from polyurethane cushion form cups against 28 mm CoCrMo heads was 14.1 ± 4.3 mg/10^6 cycles. This compared with 44.8 ± 3.4 mg/10^6 cycles for UHMWPE cups. A similar result was observed by Bigsby et al. [23] who tested 32 mm-diameter polyurethane cups against a stainless steel femoral head. In this case, the wear rates were 10.3 and 34.3 mm³/10^6 cycles for polyurethane and UHMWPE, respectively.

16.3
TriboFit® Hip System

Polyurethane was first used in humans as an acetabular component in 1960 by Charles Townley [24]. Acetabular cups made from this material were actually used 2 years before Sir John Charnley first used UHMWPE for the same indication [25]. In Townley's series of 26 patients, polyurethane failed in all cases. However, polyurethane did not cause any osteolytic reaction. The crude polyurethane used in these cases was far from being a modern bearing material, as it was prepared by mixing the pre-polymer with resin and the catalyst at the time of surgery. This was then shaped to the femoral prosthesis. Polyurethane served as both the anchoring cement for the femoral side and as the articular replacement and cement for the acetabulum. Serious efforts to further research polyurethane's possible use as a weight-bearing material did not resume until 1974 [4]. Since then, several other preclinical studies and reports have shown that a soft, compliant material such as polycarbonate-urethane may be a viable alternative to UHMWPE since it also has excellent wear resistance properties plus the ability to deform like cartilage and restore the tribologically important synovial fluid film between the articulating surfaces [5, 6, 12].

Since February 2006, polycarbonate-urethane (PCU) Bionate® 80A (DSM PTG, Berkley, CA) has been used as a cushion form acetabular component for THR (TriboFit® Hip System, Active Implants Corporation, Memphis, TN). Commercially known as the Tribofit® Acetabular Buffer™, it is an uncemented 2.7 mm-thick component which can be coupled to large diameter CoCrMo prosthetic heads. The diameter of the acetabular component ranges from 40 to 56 mm, and is available in 2 mm increments to match the diameters of the femoral head component. The Tribofit® Acetabular Buffer™ can either be implanted directly into the socket or used as a liner inserted into a specially manufactured metal shell (Fig. 16.1).

a   **b**

Fig. 16.1 (**a**) Tribofit® Acetabular Buffer™. (**b**) Tribofit® Acetabular Buffer™ in metal shell

This innovative bearing allows fluid film lubrication, and as a result, wear is minimized. Polycarbonate-urethane is a hydrophilic, biocompatible, endotoxin-resistant material and acts as a stress-absorber, transmitting loads to the subchondral bone in a physiological manner.

Preoperative planning and radiographic templating are essential for accurate and successful implantation of the TriboFit® Hip System™. The acetabulum is completely exposed and is then carefully reamed so as to remove the cartilage. A circumferential groove (Fig. 16.2a) is made in the acetabulum using the Tribofit® Groover. The Tribofit® Acetabular Buffer™ has a 2 mm peripheral ridge that snap-fits into the circumferential groove in the acetabulum (Fig. 16.2). A "spin test" is required to ensure that the Tribofit® Acetabular Buffer™ is securely (Fig. 16.2b) seated within the acetabulum. This is achieved by using special instruments to articulate the appropriate femoral head on the Tribofit® Acetabular Buffer™, using a drop of fluid for lubrication. The Tribofit® Acetabular Buffer™ component is seated properly if the femoral head spins freely.

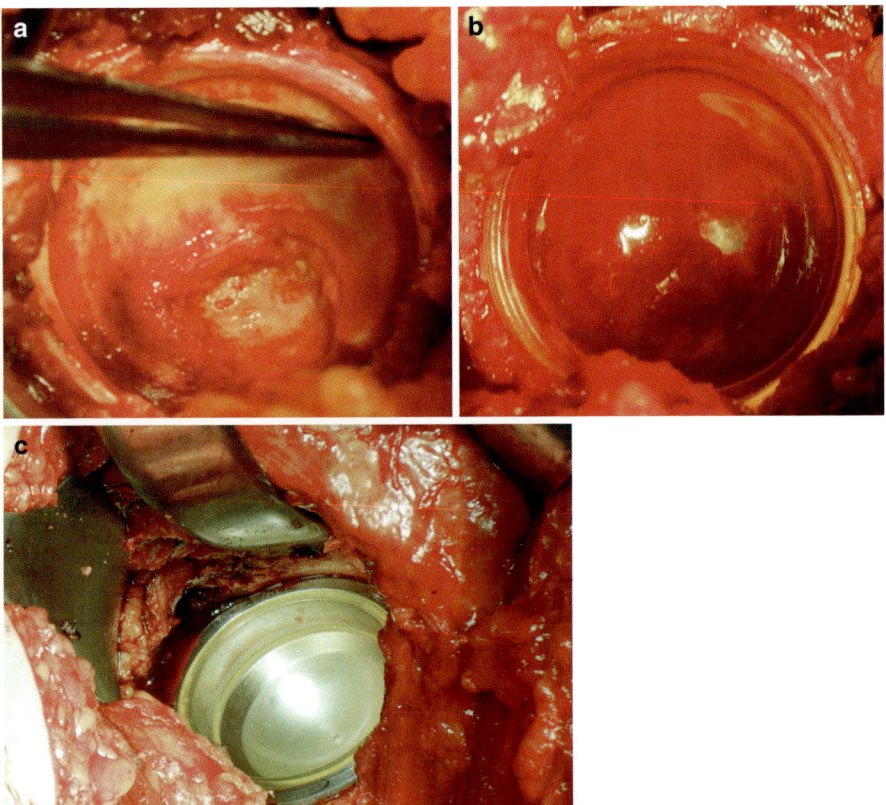

Fig. 16.2 (**a**) Acetabular exposure showing the minimal reaming and circumferential groove for implantation of the TriboFit® Acetabular Buffer™ component. (**b**) TriboFit® Acetabular Buffer™ implanted in the grooved acetabulum. (**c**) TriboFit® metal shell with the Acetabular Buffer™ component as a liner

As aforementioned, the Tribofit® Acetabular Buffer™ can also be implanted as a liner in a cementless, hydroxyapatite-coated, press-fit metal back (Fig. 16.2c) which ranges in size from 46 to 62 mm. The inner part of the shell features a 2 mm groove which corresponds to the 2 mm peripheral ridge on the Tribofit® Acetabular Buffer™. When using this configuration, standard acetabular bone reaming is required. The metal shell is press-fit into the reamed socket by using a special cup introducer. After the shell is fully seated, the cup introducer is unlocked and removed. Finally, the Tribofit® Acetabular Buffer™ is manually snapped into the grooved inner part of the shell.

16.4
Biomechanical Properties

Long-term wear experiments are an important tool with which to evaluate the potential of joint replacement devices during the stages of research and development [26, 27]. The first wear tests on the TriboFit® Acetabular Buffer™ were conducted in 2002. In this study, Fisher and Jennings recorded volumetric changes in the TriboFit® Acetabular Buffer™ over 5 million load cycles, as well as changes in the surface roughness of the TriboFit® Acetabular Buffer™ and femoral heads, as indicators of wear. The protocol followed previous studies in which one station was loaded in the vertical axis only and used as a control to compensate for geometrical changes due to creep and water absorption of the material [23, 28–30]. The findings were promising, showing a wear rate of only $2.8 \text{ mm}^3/10^6$ cycles ($\sim 3.0 \text{ mg}/10^6$ cycles). More recently, St. John [31] used a similar protocol to conduct a comparative study on the wear performance of PCU, UHMWPE and cross-linked UHMWPE buffers with similar geometrical configurations. In this study, the wear rate was measured using both gravimetric (based on the methods of Scholes et al. [30]) and particulate analysis methods. The results of these tests demonstrated a wear rate of $22.7 \text{ mg}/10^6$ cycles, which is one order of magnitude higher than that reported by Fisher and Jennings. However, under similar testing conditions, the wear rate of the TriboFit® Acetabular Buffer™ was shown to be lower than that of UHMWPE and cross-linked UHMWPE cups which had a wear rate of 80.0 and $25.4 \text{ mg}/10^6$ cycles, respectively.

In 2009, Elsner et al. [32] used a physiological simulator to evaluate the wear characteristics of the TriboFit® Acetabular Buffer™, coupled against a CoCrMo femoral head. The wear rate was evaluated over 8×10^6 cycles gravimetrically, as well as by wear particle isolation using filtration and bio-ferrography (BF). The gravimetric and BF methods showed a wear rate of $9.9–12.5 \text{ mg}/10^6$ cycles, whereas filtration resulted in a lower wear rate of 5.8 mg per million cycles. Polycarbonate-urethane demonstrated a low particle generation rate ($1–5 \times 10^6$ particles per million cycles), with the majority (96.6%) of wear particle mass lying above the biologically active range, 0.2–10 μm. Thus, PCU offers a substantial advantage over traditional bearing materials, not only in its low wear rate, but also in its osteolytic potential.

The optimal wear characteristics of this material are associated with an ideal lubrication. Indeed, the thickness of the fluid film which separates the articular surfaces is the same as in the normal hip joint and is much thicker than other bearing technologies such as metal-on-UHMWPE, COC and MOM (Fig. 16.3).

16

Fig. 16.3 Bar chart showing the fluid film thickness present in normal joints and in those replaced by various bearing types. The fluid film thickness in metal-on-PCU bearings is comparable to that in cartilage-on-cartilage

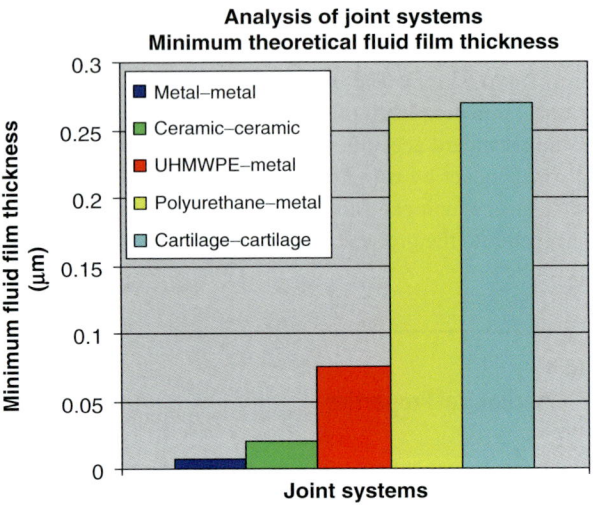

16.5
Clinical Experience

16.5.1
Femoral Neck Fracture Series

Although the incidence of femoral neck fractures is increasing, the treatment of elderly patients with displaced femoral neck fractures is still a challenge [33]. Despite the large number of cases, the outcomes are not as good as with other hip arthroplasty indications. The preferred treatment options for displaced femoral neck fractures are hemiarthroplasty and THR. The advantages of hemiarthroplasty include no acetabular bone reaming, a short surgical time and a reduced risk of dislocation. One major drawback however is reduced implant longevity. Implant longevity is better with total hip arthroplasty but other drawbacks exist such as a higher risk of dislocation and increased surgical time due to the need for acetabular bone reaming [34–38].

The use of the TriboFit® Hip System, using the TriboFit® Acetabular Buffer™ alone, has the potential advantage of combining all the benefits of hemiarthroplasty and THR, without any of the original drawbacks. Unlike other arthroplasty procedures, this includes the use of a prosthetic head with the same dimensions as the patient's original femoral head.

As of September 2010, this system has been used clinically for nearly 5 years by over 100 surgeons to treat femoral neck fractures in 714 patients. The average follow-up for all the patients was 12.1 months. Out of these 714 patients, data is available on 394. The series comprises 24% males and 76% females, with a mean age of 80. The mean operating time was 88 min (75–171 min). Rehabilitation was fast and good hip function was resumed in the majority of patients. There were 13 major complications: two infections (0.5%), one of which was a deep infection which necessitated removal of the implant. The second infection was superficial and was treated by debridement and antibiotics, without removal of the implant. There were seven dislocations (1.8%) which occurred within 4 months after surgery (five of the dislocations were surgically treated with removal of the TriboFit® Acetabular Buffer™ and implantation of a different system, one was treated with surgical reduction and one with closed reduction), two dislodgements of the TriboFit® Acetabular Buffer™ (0.5%), both of which were revised using a different arthroplasty system. One dislodgement occurred almost immediately after surgery but was only revised 9 months afterwards. The second patient who had a dislodgement initially complained of a noisy hip which then subsided. After a few years of good function and no pain, the patient became symptomatic and was revised 44 months after surgery. There was another patient who complained of a noisy hip. This patient was initially asymptomatic but was revised 10.5 months after surgery when pain developed. In total, to date, 12 patients (3.0%) have been revised. In 11 of these patients (2.8%), the TriboFit® Acetabular Buffer™ was removed.

As reported by Siebert et al. and Wippermann et al. [39, 40], analyses of the retrievals confirmed the optimal wear properties of this new bearing. The synovial fluid was normal and there were no adverse tissue reactions. There was minimal wear of the articulating surface, which was actually less than expected based on the in vitro results. Backside wear was more severe.

The results from this series are similar to those reported in the literature for femoral neck fractures treated with other arthroplasty types at short-term. A great advantage of this surgical technique is the low rate of dislocation, which is similar to the dislocation rate reported in the literature for femoral neck fractures treated with hemiarthroplasty, and lower than the dislocation rate reported for femoral neck fractures treated with THR [41]. We believe the low dislocation rate observed with the TriboFit® Hip System is due to the use of large diameter metal heads which replicate the size and diameter of the patient's original head. This is particularly important with the elderly population who do not tolerate any change in their hip joint anatomy (Fig. 16.4).

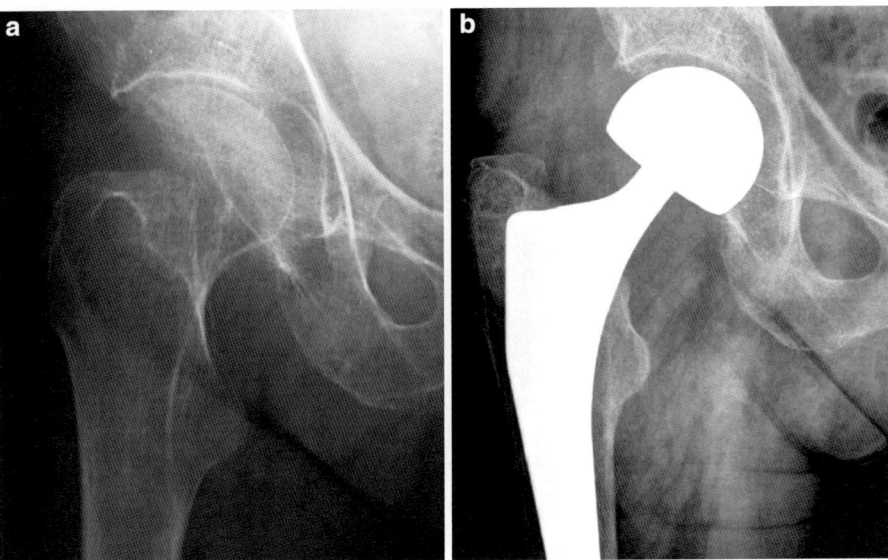

Fig. 16.4 (**a**) An 87-year-old female with a femoral neck fracture. (**b**) Three years after treatment with the Tribofit® Hip system

16.5.2
Osteoarthritis

When degenerative changes due to osteoarthritis affect the geometry of the acetabulum, the Tribofit® Acetabular Buffer™ component alone is not indicated. In deformed sockets, standard acetabular bone reaming is needed in order to regain the previous socket sphericity. By implanting the Tribofit® Acetabular Buffer™ directly into the reamed acetabulum, the backside would be exposed to contact with spiky bone and this would potentially lead to increased backside wear. In such cases, the TriboFit® Acetabular Buffer™ can be used as an articulating bearing surface inside a press-fit metal back made from CoCrMo and coated with hydroxyapatite in order to optimize osteointegration.

As of September 2010, the TriboFit® Hip System has been implanted with a metal back in 157 patients treated with THR and four patients treated with hip resurfacing. The average follow-up time is 12.1 months (range 0–32). The clinical outcomes were good and rehabilitation was fast and uncomplicated. The range of motion was normal and the patients reported that their hip felt normal. There were no dislocations, dislodgements, or infections. One total hip case has been revised because of loosening of the acetabular component caused by an intraoperative acetabular fracture. At the latest follow-up, X-rays showed good bone-implant contact with no signs of osteolysis or bone rarefaction (Fig. 16.5).

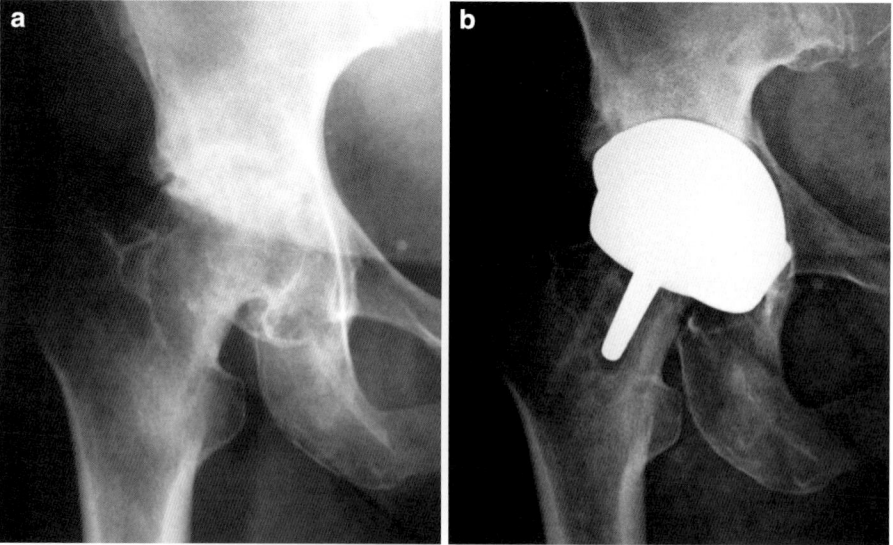

Fig. 16.5 (**a**) A 63-year-old female with osteoarthritis. (**b**) One year after treatment with the Tribofit® Acetabular Buffer™ in a metal shell

16.6
Discussion

With polyethylene, wear leading to perimplant osteolysis and mechanical loosening is a major concern, particularly in the younger, more active patient population [42–47]. Hard-on-hard bearing surfaces (COC and MOM) offer the potential to decrease the incidence of revision surgery caused by wear but they possess other disadvantages.

Polycarbonate-urethane is a new bearing option for the prosthetic treatment of hip joint diseases. This new bearing is highly resistant to wear and fatigue, with ideal lubrication and a more physiological contact stress distribution. The size and morphology of the debris created from this innovative bearing is less reactive than that produced by other bearing technologies [48, 49]. Another major advantage is the use of large diameter heads, which are crucial for optimizing joint stability and range of motion [50, 51].

The use of the Tribofit® Acetabular Buffer™ alone has great potential for the treatment of femoral neck fracture patients as it combines the advantages of hemiarthroplasty and THR without any of their original drawbacks. However, surgeons should bear in mind that implanting a cushion form acetabular component directly into an unreamed acetabulum requires accurate technique, a high degree of surgical skill, and adequate bone quality. This procedure cannot be used in patients with severe osteoporosis, osteoarthritic changes or subchondral cysts. We believe that some of the complications observed in our femoral neck fracture group were caused by inaccurate surgical technique or incorrect indications. We expect even better clinical outcomes and a lower incidence of complications when the surgeon's learning curve is complete.

In the treatment of osteoarthritis patients, standard acetabular bone reaming is required and the Tribofit® Acetabular Buffer™ alone cannot be used because of the risk of backside wear. In these cases, optimal results can be achieved by using an uncemented hydroxyapatite-coated metal shell which features the Tribofit® Acetabular Buffer™ as a liner. We believe that this bearing configuration has unique advantages over other bearing technologies in improving both clinical outcomes and implant longevity in patients treated with THR. In conclusion, the Tribofit® Hip System is a promising bearing technology which has already been validated in a clinical setting.

References

1. Callaghan, J.J., Templeton, J.E., Liu, S.S., Pedersen, D.R., Goetz, D.D., Sullivan, P.M., Leinen, J.A., Johnston, R.C.: Results of Charnley total hip arthroplasty at a minimum of thirty years. A concise follow-up of a previous report. J. Bone Joint Surg. Am. **86**(4), 690–695 (2004)
2. Kusaba, A., Kuroki, Y.: Femoral component wear in retrieved hip prostheses. J. Bone Joint Surg. Br. **79B**, 331–336 (1997)
3. Jasty, M.J., Floyd, W.E., Schiller, A.L., Goldring, S.R., Harris, W.H.: Localised osteolysis in stable, non-septic total hip replacement. J. Bone Joint Surg. Am. **68A**, 912–919 (1986)

4. Howie, D.W., Haynes, D.R., Rogers, S.D., McGee, M.A., Pearcy, M.J.: The response to particulate debris. Orthop. Clin. N. Am. **24**, 571–581 (1993)
5. Atkinson, J.R., Dowson, D., Isaac, J.H., Wroblewski, B.M.: Laboratory wear tests and clinical observations of the penetration of femoral heads into acetabular cups in total replacement hip joints. Part III: the measurement of internal volume changes in explanted Charnley sockets after 2-16 years in vivo and the determination of wear factors. Wear **104**, 225–244 (1985)
6. Devane, P.A., Bourne, R.B., Rorabeck, C.H., MacDonald, S., Robinson, E.J.: Measurement of polyethylene wear in metal-backed acetabular cups. Clin. Orthop. Relat. Res. **319**, 317–326 (1995)
7. Schmalzried, T.P., Jasty, M., Harris, W.H.: Periprosthetic bone loss in total hip arthroplasty: polyethylene wear debris and the concept of the effective joint space. J. Bone Joint Surg. Am. **74-A**, 849–863 (1992)
8. Currier, B.H., Currier, J.H., Mayor, M.B., Lyford, K.A., Collier, J.P., Van Citters, D.W.: Evaluation of oxidation and fatigue damage of retrieved crossfire polyethylene acetabular cups. J. Bone Joint Surg. Am. **89**(9), 2023–2029 (2007)
9. Tower, S., Currier, J.H., Currier, B.H., Lyford, K.A., Van Citters, D.W., Mayor, M.B.: Rim cracking of the cross-linked longevity polyethylene acetabular liner after total hip arthroplasty. J. Bone Joint Surg. Am. **89**, 2212–2217 (2007)
10. Jarrett, C.A., Ranawat, A.S., Bruzzone, M., Blum, Y.C., Rodriguez, J.A., Ranawat, C.S.: The squeaking hip: a phenomenon of ceramic-on-ceramic total hip arthroplasty. J. Bone Joint Surg. Am. **91**(6), 1344–1349 (2009)
11. Wagner, H., Wagner, M.: German clinical results with Metasul bearings. In: Rieker, C., Windler, M., Wyss, U. (eds.) Metasul: A Metal on Metal Bearing. Hans Huber, Bern (1999)
12. Delaunay, C., Petit, I., Learmonth, I.D., Oger, P., Vendittoli, P.A.: Metal-on-metal bearings total hip arthroplasty: the cobalt and chromium ions release concern. Orthop. Traumatol. Surg. Res. **8**(2), 87–98 (2010)
13. Unsworth, A., Dowson, D., Wright, V.: Some new evidence on human joint lubrication. Ann. Rheum. Dis. **34**(4), 277–285 (1975)
14. Scholes, S.C., Unsworth, A., Blamey, J.M., Burgess, I.C., Jones, E., Smith, N.: Design aspects of compliant soft layer bearings for an experimental hip prosthesis. Proc. Inst. Mech. Eng. H **219**(2), 79–87 (2005)
15. Dowson, D.: Basic tribology. In: Dowson, D., Wright, V. (eds.) An Introduction to the Biomechanics of Joints and Joint Replacement. Mechanical Engineering Publications Limited, London (1981)
16. Unsworth, A., Roberts, B., Thompson, J.C.: The application of soft-layer lubrication to hip prostheses. J. Bone Joint Surg. **63B**, 297 (1981)
17. Unsworth, A., Dowson, D., Wright, V.: The frictional behaviour of human synovial joints – Part 1: natural joints. Trans. ASME J. Lubric. Technol. **97**(3), 369–376 (1975)
18. Dowson, D., Jin, Z.M.: Micro-elastohydrodynamic lubrication of synovial joints. Eng. Med. **15**(2), 63–65 (1986)
19. Unsworth, A., Pearcy, M.J., White, E.F.T., White, G.: Soft layer lubrication of artificial hip joints. In: Proceedings of IMechE International Conference of Fifty Years on, London, 1987, pp. 715–724. Mechanical Engineering Publications, London (1987)
20. Burgess, I.C.: Tribological and mechanical properties of compliant bearings in total joint replacement. Dissertation, University of Durham (1997)
21. Yao, J.Q., Unsworth, A.: Asperity lubrication of human joints. Proc. Inst. Mech. Eng. H **207**, 245–254 (1993)
22. Smith, S.L., Ash, H.E., Unsworth, A.: A tribological study of UHMWPE acetabular cups and polyurethane compliant layer acetabular cups. J. Biomed. Mater. Res. **813B**, 710–716 (2000)
23. Bigsby, R.J.A., Auger, D.D., Jin, Z.M., Dowson, D., Hardaker, C.S., Fisher, J.: A comparative tribological study of the wear of composite cushion cups in a physiological hip joint simulator. J. Biomech. **31**, 363–369 (1998)

24. Townley, C.O.: Hemi and total articular replacement arthroplasty of the hip with the fixed femoral cup. Orthop. Clin. North Am. **13**, 869–894 (1982)
25. Charnley, J.: The long-term results of low-friction arthroplasty of the hip performed as a primary intervention. J. Bone Joint Surg. Br. **54**(1), 61–76 (1972)
26. Scholes, S.C., Inman, I.A., Unsworth, A., Jones, E.: Tribological assessment of a flexible carbon-fibre reinforced poly(ether–ether–ketone) acetabular cup articulating against an alumina femoral head. N. Engl. J. Med. **222**(H), 273–283 (2008)
27. Scholes, S.C., Unsworth, A.: Wear studies on the likely performance of CFR-PEEK/CoCrMo for use as artificial joint bearing materials. J. Mater. Sci. Mater. Med. **20**, 163–170 (2009)
28. Dowson, D., Jobbins, B., Seyed-Harraf, A.: An evaluation of the penetration of ceramic femoral heads into polyethylene acetabular cups. Wear **880**, 889 (1993)
29. Bigsby, R.J.A., Hardaker, C.S., Fisher, J.: Wear of ultra-high molecular weight polyethylene acetabular cups in a physiological hip joint simulator in the anatomical position using bovine serum as a lubricant. Proc. Inst. Mech. Eng. H **211**(H), 265–269 (1997)
30. Scholes, S.C., Unsworth, A., Jones, E.: Polyurethane unicondylar knee prostheses: simulator wear tests and lubrication studies. Phys. Med. Biol. **52**, 197–212 (2007)
31. St. John, K.R., Gupta, M.N.: Wear resistance of polycarbonate urethane acetabular cups as compared to cross-linked ultra high molecular weight polyethylene. Society for Biomaterials Annual Meeting (2010)
32. Elsner, J.J., Mezape, Y., Hakshur, K., Shemesh, M., Linder-Ganz, E., Shterling, A., Eliaz, N.: Wear rate evaluation of a novel polycarbonate-urethane cushion form bearing for artificial hip joints. Acta Biomater. **6**(12), 4698–4707 (2010)
33. Steenbrugge, F., Govaers, K., Van Nieuwenhuyse, W., Van Overshelde, J.: Displaced femoral neck fractures in adult and elderly patients, pitfalls in treatment options: internal fixation or replacement? Acta Orthop. Traumatol. Turc. **34**, 430–433 (2000)
34. Sierra, R.J., Cabanela, M.E.: Conversion of failed hemiarthroplasties after femoral neck fractures. Clin. Orthop. Relat. Res. **399**, 129–139 (2002)
35. Gilbert, M., Capozzi, J.: Unipolar or bipolar prosthesis for the displaced intracapsular hip fracture? An unanswered question. Clin. Orthop. Relat. Res. **353**, 81–85 (1998)
36. Lachiewicz, P.F., Soileau, E.S.: Stability of arthroplasty in patients 75 years or older. Clin. Orthop. Relat. Res. **405**, 65–69 (2002)
37. Steinberg, M., Corces, A., Fallon, M.: Acetabular involvement in osteonecrosis of the femoral head. J. Bone Joint Surg. Br. **81**, 60–65 (1999)
38. Lester, D.K., Wertenbruch, J.M., Piatkowski, A.M.: Degenerative changes in normal femoral heads in the elderly. J. Arthroplasty **14**(2), 200–203 (1999)
39. Siebert, W.E., Mai, S., Kurtz, S.: Retrieval analysis of a polycarbonate-urethane acetabular cup: a case report. J. Long Term Eff. Med. Implants **18**(1), 69–74 (2008)
40. Wippermann, B., Kurtz, S., Hallab, N., Treharne, R.: Explantation and analysis of the first retrieved human acetabular cup made of polycarbonateurethane: a case report. J. Long Term Eff. Med. Implants **18**(1), 75–83 (2008)
41. Bhandari, M., Devereaux, P.J., Swiontkowski, M.F., Tornetta III, P., Obremskey, W., Koval, K.J., Nork, S., Sprague, S., Schemitsch, E.H., Guyatt, G.H.: Internal fixation compared with arthroplasty for displaced fractures of the femoral neck. A meta-analysis. J. Bone Joint Surg. Am. **85-A**, 1673–1681 (2003)
42. Dunstan, E., Ladon, D., Whittingham-Jones, P., et al.: Chromosomal aberrations in the peripheral blood of patients with metal-on-metal bearings. J. Bone Joint Surg. Am. **90**(3), 517–522 (2008)
43. Murphy, S.B., Ecker, T.M., Tannast, M.: Two- to 9-year clinical results of alumina ceramic-on-ceramic THA. Clin. Orthop. Relat. Res. **453**, 97–102 (2006)
44. Willmann, G.: Ceramic femoral head retrieval data. Clin. Orthop. Relat. Res. **379**, 22–28 (2000)
45. Yang, C.C., Kim, R.H., Dennis, D.A.: The squeaking hip: a cause for concern-disagrees. Orthopaedics **30**(9), 739–742 (2007)

46. Ranawat, A.S., Ranawat, C.S.: The squeaking hip: a cause for concern-agrees. Orthopaedics **30**(9), 738–743 (2007)
47. Taylor, S., Manley, M., Sutton, K.: The role of stripe wear in causing acoustic emissions from alumina ceramic-on-ceramic bearings. J. Arthroplasty **22**, 47–51 (2007)
48. Smith, R.A., Hallab, N.J.: In vitro macrophage response to polyethylene and polycarbonate-urethane particles. J. Biomed. Mater. Res. A **93**(1), 347–355 (2010)
49. Smith, R.A., Maghsoodpour, A., Hallab, N.J.: In vivo response to cross-linked polyethylene and polycarbonate-urethane particles. J. Biomed. Mater. Res. A **93**(1), 227–234 (2010)
50. Burroughs, B.R., Hallstrom, B., Golladay, G.J., Hoeffel, D., Harris, W.H.: Range of motion and stability in total hip arthroplasty with 28, 32, 38 and 44 mm femoral heads. J. Arthroplasty **20**(1), 11–19 (2005)
51. Berry, D.J., von Klnock, M., Schleck, C.D., Harmsen, W.S.: Effect of femoral head diameter and operative approach on risk of dislocation after primary total hip arthroplasty. J. Bone Joint Surg. Am. **87**, 2456–2463 (2005)

A Novel Model to Predict Wear in an Uncemented Hip Replacement with a Ceramic on Polyethylene Bearing

17

Simon Boyle, Peter Loughenbury, Phil Deacon, and Richard M. Hall

17.1
Introduction

Monitoring of implants is becoming necessary to identify early signs of failure, and guide the frequency of follow-up review. This is particularly important in the growing population of young and active patients undergoing total hip replacement (THR). The most common mode of THR failure is that of aseptic loosening due to osteolysis, a process that is strongly associated with the generation of bioactive wear debris. The presence of osteolysis, excessive wear and other warning signs of failure can be determined with conventional radiography. We identified a population of young patients undergoing THR using the Furlong system (with a metal backed cup comprising a ceramic on polyethylene bearing) in which a measure of early failure would be useful in guiding follow-up review. The purpose of this initial study was to create a model to predict wear in this patient cohort that could then be validated in a larger patient group, which may involve prostheses of other types.

S. Boyle (✉)
Department of Trauma and Orthopaedics, York Teaching Hospital,
Wigginto Road, YO31 8HE York, UK and
University of Leeds, Leeds, UK
e-mail: siboyle@me.com

P. Loughenbury
SpR Trauma and Orthopaedics, Yorkshire Deanery, UK and
School of Mechanical Engineering, University of Leeds, Leeds, UK
e-mail: prloughenbury@hotmail.com

R.M. Hall
Professor of Spinal Biomechanics
School of Mechanical Engineering, University of Leeds, Leeds, UK
e-mail: r.m.hall@leeds.ac.uk

P. Deacon
Consultant Orthopaedic Surgeon, Department of Trauma and Orthopaedics,
Pinderfields General Hospital, Aberford Road, Wakefield WF1 4DG, UK

K. Knahr (ed.), *Tribology in Total Hip Arthroplasty*,
DOI: 10.1007/978-3-642-19429-0_17, © 2011 EFORT

17.2
Methods

Between March 1993 and April 2004, non-consecutive uncemented total hip replacements were performed by a single surgeon (PD) using a ceramic on polyethylene bearing combination. Fifty-nine hips in 43 patients were randomly selected from clinic attendance for clinical and radiographic follow-up. There were 19 male (27 hips) and 24 female patients (32 hips) and the mean age at the time of surgery was 53 (range 25–71). Seventeen patients had undergone staged bilateral procedures. Fifty-five procedures were carried out for primary osteoarthritis. The other indications for surgery were avascular necrosis, post-traumatic osteoarthritis after a hip fracture, developmental dysplasia of the hip, and revision of a failed cemented THR.

A Furlong hydroxyapatite-coated femoral stem (Joint Replacement Instrumentation [JRI], London, UK) was used in all patients. Pre-operative templating and intra-operative reaming determined the size of femoral component used. 28 mm alumina femoral heads were used in all patients. Two different types of acetabular component were used, threaded and hemispherical, both of which are HA coated. The threaded cup comprises a parallel-sided screw thread, which is pre-tapped prior to insertion. The hemispherical cup engages with an interference fit and contains multiple screw holes for drilling and insertion of 6.5 mm screws to aid in early stability and fixation. Pre-operative templating and intra-operative reaming determined the size of both types of acetabular component. Ultra high molecular weight polyethylene (UHMWPE) liners with a minimum thickness of 8 mm were used. Patients were positioned supine and a lateral Hardinge approach was used to gain access to the hip joint. Follow-up occurred at 6 weeks, 6 months and 1 year following the procedure, and annually thereafter.

Radiographic assessment was performed at each follow-up appointment but only the post-operative and most recent x-rays were examined in this study. Standard AP films of the pelvis were taken with the patient supine and centred on the pubic symphysis at a distance of 1 m. Radiographs were scanned with a Vidar VXR-12 film digitiser with a resolution of 300 DPI on a $4,000 \times 4,000$ pixel matrix. The imaging software used was isolutionlite version 7.3 (Image and Microscope technology, Coquitlam, BC, Canada). All measurements were performed by one observer (SB). Heterotrophic Ossification (HO) was assessed and graded according to the Brooker [1] classification. Osteolysis was assessed by examining Gruen's zones [2] around the femoral component and DeLee and Charnley's [3] zones around the acetabular component. Osteolysis was defined as any area of progressive lucency surrounding the prosthesis and measuring >2 mm in its greatest diameter. Loosening of either the acetabulum or the femur was defined as the presence of radiolucent lines (extending at least 50% of the particular zone [4]) at the interface between the component and bone that had not been seen on the first post-operative film.

The presence of a pedestal, a shelf of endosteal bone either partially or completely bridging the intra-medullary canal at the level of the tip of the prosthesis, was noted. Collar-calcar contact (visible contact between the collar of the prosthesis and the calcar seen on x-ray) and calcar resorption (>2 mm resorption of bone at the level of the calcar that did not extend as a linear lucency down the bone implant interface) were also recorded.

Stress Shielding was noted; defined as a decrease in the apparent bone density reflected in a relative osteopenia as compared to the surrounding bone and post-operative films. Stem alignment (angular deviation of the stem from the long axis of the femur) and acetabular inclination (determined by using the inter-teardrop line as the horizontal reference in the pelvis [5]) was measured three times for both x-rays and the mean taken.

Wear was measured using the technique described by Livermore [6] using plain x-rays illuminated by a light box in a darkened room and a set of digital callipers (Digimatic 500-191U, Mitutoyo Ltd, Hampshire, UK) which had a minimum resolution of 0.01 mm. The centre of the femoral head was located by using a transparent template of concentric circles and marking the centre of the head on the x-ray. The radius of the head (R_f) was then measured at 3 points around the head to ensure that this centre had been localised as closely as possible. If the difference in any of these measurements was greater than 0.1 mm then the template was repositioned and the process was repeated. The calliper was then used to measure the shortest distance from the femoral head centre to the edge of the metal backed acetabulum. Three measurements were taken and the mean value used. This was recorded for both the post-operative x-ray (A_1) and the latest film (A_2). The difference in the two measurements was determined to be the distance of migration of the femoral head in the coronal plane. This depth was then converted into millimetres by calibrating it with the known radius of the femoral head using the equation below.

$$\text{Migration of the femoral head} = (R_f/14)(A_1 - A_2)$$

17.2.1
Statistical Analysis

Data analysis was performed using Statistical Package for Social Sciences (SPSS) version 15.0, 2006. Descriptive statistics were used to describe demographic data and the wear related variables including the mean, range and standard deviation. All data variables were first examined with a scatterplot and histogram with a distribution curve to examine for normality of distribution. Where associations were sought between variables that were normally distributed, Pearson's correlation co-efficient (ρ) was used to measure the strength of any association and its significance (p value). Where these distributions were not normal, Spearman's correlation co-efficient (r_s) and significance testing were applied. When testing the significance of the difference between two means or medians in paired data sets that were not normally distributed, then Wilcoxon's signed rank test was used. The Mann Whitney U test was used if the data sets were non-parametric and unpaired. To assess for a potential cause and effect relationship between the dependent variables a scatter-plot was constructed for the two variables and this was examined for linearity. In instances where the dependent variable was not normally distributed, an attempt to normalise this data was performed by taking logs or squaring the data. Univariate linear regression analysis was then performed for each independent variable and the main dependent variables to give an R^2 value, slope co-efficient, p value and confidence intervals. Independent variables with a p value <0.2 were retained and the most statistically significant variable (the variable with the lowest p value) was used as the starting point for a multivariate linear regression model. Variables were then added using forward selection. The next most significant variable was added in to the model and the effect on the R^2 value and its significance noted. Where a subsequent independent variable was found to improve the R^2 value for the model and was statistically significant, it was retained in the model. If a variable caused a noticeable change in an existing variable co-efficient then this variable was retained as a potential confounder. This process was repeated until the final model was created. Linear regression analysis was used in this way to define the true contribution of an independent variable to the proposed model constructed to predict the dependent variable. Significance values were set as $p < 0.05$ and 95% confidence intervals were calculated where appropriate.

17.3
Results

The average age at time of follow-up was 53 (34–76). In males, the mean age was 57 and in females this was 50. The mean length of time between the postoperative x-ray and the latest follow-up x-ray was 53 months (11–162) and was designated the implant period. There was no sign of femoral stem migration or instability. Calcar-collar contact was achieved in 51 out of 59 prostheses on the initial postoperative x-ray and calcar resorption was seen in 12 out of 59. Nine out of 59 hips demonstrated the presence of a pedestal at the tip of the prosthesis. A statistically significant correlation was found between femoral alignment and calcar resorption (Spearman's correlation co-efficient −0.301, $p = 0.021$). Other variables tested included time since implantation, total wear, log wear, annual wear, volumetric wear, type of cup and presence of osteolysis in the acetabulum. None of these resulted in a statistically significant association. Fifty-one out of the 59 procedures (26 male, 25 female) used a hemispherical fully coated cup. The remaining eight (one male, seven female) received threaded Furlong HA coated cups. HO was seen in 24 out of the 59 hips. According to the Brooker [1] classification, 17 were grade I, 5 grade II and two were grade III. The angle the acetabulum tended to the pelvis increased from a mean of 40.8° on the post op film to 41.9° in the latest film (mean difference of 1.1°, $p < 0.001$). Eleven cups were measured to be in a more horizontal (closed position) at the latest follow-up as compared to the post op x-ray whilst two were unchanged and 46 were more vertical (open).

17.3.1
Wear Measurements

Total wear was recorded as the maximal migration of the femoral head centre from its initial position on the postoperative film to the latest x-ray. Wear measurements were not possible in two hips due to the quality of the x-rays taken leaving 57 out of the 59 hips available for wear examination. Fifteen out of the 57 total wear measurements were negative values, which can arise from different projections between the most recent and post-operative x-rays, operator error or subluxation of the head inside the cup. The median value of the penetration was 0.11 mm with an interquartile range of −0.1 mm to 0.24 mm and a maximum value of 4.48 mm. Initial analysis demonstrated that there was a highly significant correlation between penetration depth versus implant period (Spearman, $r=0.60$, $P<0.0001$) and cup inclination (Spearman, $r=0.70$, $P<0.0001$).

Volumetric wear was calculated using the formula $\pi r^2 h$ where h was the linear penetration or total wear depth. The median volumetric wear for all the hips was 70.6 mm^3 with an interquartile range of 0–147.8 mm^3 and a maximum value of 2759 mm^3. There was no association found between total volumetric wear and acetabular osteolysis or calcar resorption.

Acetabular osteolysis was seen in five cases only. The osteolysis seen was focal and was found in zone I in four cases and zone III in one case. None of the implants were radiologically unstable. Statistically significant correlations were found between acetabular osteolysis and time since implantation ($r_s=0.285$, $p=0.032$), patient age ($r_s=-0.332$, $p=0.014$) and acetabular alignment ($r_s=0.281$, $p=0.031$).

Intra-observer variability was tested with ten randomly selected x-rays re-measured at an interval of 1 month. The femoral head radius and the total wear were recorded and compared to the measurements previously taken. Measurement variability was consistent throughout a range of wear and femoral head sizes. For total wear the mean standard error was 0.061 mm, the mean standard deviation was 0.086 mm and intra-class correlation co-efficient (ICC) was 0.998. For femoral head measurements, mean standard error was 0.059 mm, mean standard deviation was 0.083 mm and ICC was 0.992.

17.3.2
Formulation of a Predictive Model

A scatter plot of total wear against time since implantation (Fig. 17.1) shows a relationship between total wear and time – as time since implantation increases so does total wear, although we can see that there are a number of outliers which have large volumetric wear at higher values of implant period. The proportion of explained to total variance for this linear regression model was determined to be $R^2=0.258$ ($P=0.0002$).

This provides the following model:

$$\text{Total wear} = 0.01*(\text{implant time}) - 0.18$$

Applying a forward or stepwise linear regression analysis including all the other variables measured did not reveal any other statistically significant relationships with total wear other than time since implantation. However, if the log is taken of total wear and the linear regression model reapplied then the spread of the residuals gives a more normal distribution. A new model incorporating log Total Wear as the dependant variable gives an R^2 value of 0.390 ($p<0.001$) and a correlation co-efficient of 0.625 ($p<0.001$). The predictive capacity of the model improves to 39% (Fig. 17.2).

This provides a new model equation

$$\text{Log of total wear} = 0.017(\text{Time since implantation in months}) - 2.647$$

17

Fig. 17.1 Scatter plot of total wear against time since implantation

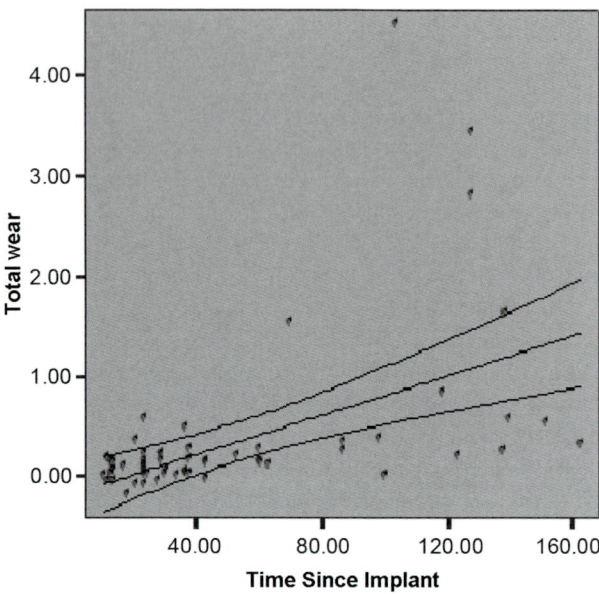

Fig. 17.2 Scatter plot of log total wear against time since implantation

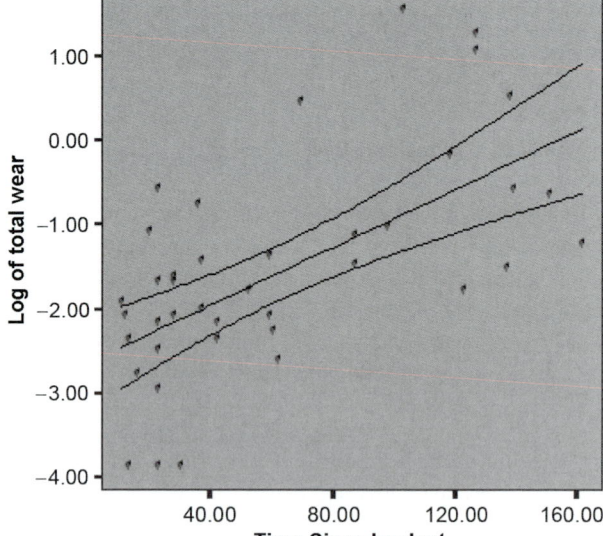

In order to include cup type, sex and side in the predictive model these variables were assigned values:

- CupCode (Hemispherical = 1; Threaded = 2)
- SexCode (Female = 1; Male = 2)
- SideCode (Left = 1; Right = 2)

A Forward Selection process was used to improve the regression model using the following relevant variables – time since implantation, type of cup, sex, side, patient age, pedestal formation, acetabular inclination and femoral alignment. Using this process, the final model now includes the variables assigned to cup type and sex as being statistically significant. The explanatory power of the model is finally improved to an R^2 value of 0.543. Taking logs of time or using time [2] did not improve the model. Applying stepwise multiple linear regressions to the sub-groups of different cup types or sex also failed to improve the model. The same results were achieved using a computerized stepwise multivariate regression process.

The final model appears as:

$$\text{Log Total wear} = (0.012 \times \text{time since implantation}) + (0.946 \times \text{CupCode}) - (0.647 \times \text{SexCode}) - 2.537$$

Therefore,

$$\text{Total wear} = 10^{(0.012 \times \text{time since implantation}) + (0.946 \times \text{CupCode}) \_ (0.647 \times \text{SexCode}) - 2.537}$$

17.4
Discussion

Identification of whether or not a prosthesis is performing poorly is important because it can guide the frequency of follow-up review. Where there are higher levels of wear and an increased risk of osteolysis a patient will need to be seen more regularly than those without. Currently, the methods of assessment include a detailed history and examination of the patient encompassing scoring systems, followed by a radiological assessment. This project focussed on the measurement of wear in a ceramic on polyethylene bearing combination in an uncemented hip prosthesis and examining any factors that may contribute to an increase in its wear given that wear debris is the main causative factor in initiating periprosthetic osteolysis. This was then progressed to construct a model to predict the amount of wear and therefore wear rate in this THR prosthesis.

This study focussed on younger patients, with a sample mean age of 53, chosen through the use of a ceramic on polyethylene bearing. One patient at the age of 71 was given the same bearing due to an excellent preoperative fitness and activity level. All patients included in the study were operated on and remained under the care of a single senior surgeon (PD) and this helped to remove any variation in wear results that may be due to surgical technique or differences in peri-operative management. Direct patient selection was determined by convenience sampling from outpatient follow-up over the period of the study. This method unfortunately introduces a risk of bias, as the sample may not be an accurate representation of all the ceramic on polyethylene hips performed by this surgeon.

A two-dimensional manual Livermore technique was used to measure wear in this study. Automated edge detection methods were considered but were not suitable as the metal backed cup meant that there would be no significant drop in the grey scale profile at the edge of the femoral head. Computerised methods have been used in previous studies but provide no better accuracy than manual method [7]. Three-dimensional analysis of wear can improve accuracy by up to 10%, although at the expense of repeatability, though it is widely accepted that two-dimensional analysis of an AP radiograph is an acceptable method of measuring wear without the need for a lateral image [8–11]. Overall, the manual Livermore technique provided a quick and efficient method of measuring wear, with excellent reported accuracy [6]. This made the technique ideal for use in a clinical setting and for use with a predictive model.

In situations with very low rates of wear, the accuracy of a technique may be close to the total wear or wear rate. The resolution of the final wear measurements in this study was examined by repeating the measurements on ten randomly selected x-rays and revealed a mean standard error of 0.061 (SD 0.086). This resolution is of the same order of magnitude as many of the wear results and reflects the difficulty in making accurate and precise measurements in this bearing couple. A further source of error lies in the measurement of the ceramic femoral head radius used to calibrate the wear depth (mean standard error 0.059; SD 0.083) and in the resolution of the digital callipers (0.01 mm), again the same order of magnitude as the wear results. Finally, in our assessment of wear, we have assumed that femoral head penetration occurs purely due to the wear of polyethylene from its articulation with the femoral head. This does not take into account the effects of settling of the liner in the cup and creep and this may overestimate the true amount of wear.

17.4.1
Wear Prediction

This project aimed to produce a model that could predict wear and be used easily in a day-to-day clinic environment, serving to highlight any poorly performing prostheses. The clinical benefit of this would be in its use as a comparison to an expected outcome. This would then identify a prosthesis that was wearing at a higher rate than expected so this patient could be monitored more closely. It would also enable the surgeon to analyse any particular cause for the increased wear such as a change in technique or a change in the design of the implant. If a trend were developing then this would give feedback to the surgeon that a change may be necessary or that previous changes have been detrimental.

To create the predictive model for total wear a linear regression analysis was used. Use of the raw data alone produced a poor predictive model, while introducing $\log_{10}$ values improved the explanatory power of the model by over 50% (R^2 increased from 0.258 to 0.390). Forward selection using all relevant independent variables then improved the model further to an R^2 value of 0.543, which had now incorporated sex and cup type into the model. The final model could easily be used in a clinical setting using the variables of time since implantation, sex and cup type to predict total wear and the time frame for liner breakthrough. The R^2 value in the final model dictates that the variations in wear measurement can be explained in 54.3% of hips. The use of logarithms in the model makes this less intuitive but it significantly improved the predictive power of the model for this data set (patient cohort). This compares favorably with R^2 values published by Dowd et al. [8] using predictive models from early wear data. However, in 45.7% of hips the model will not account for wear variance. It is likely that larger patient numbers would improve the predictive power of the model beyond 54.3%.

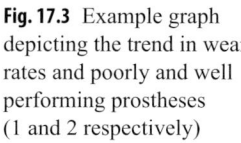

Fig. 17.3 Example graph depicting the trend in wear rates and poorly and well performing prostheses (1 and 2 respectively)

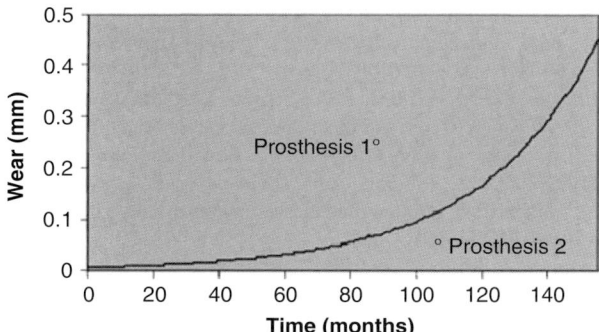

Total wear can be measured and compared to the expected wear from this equation or from a pre-printed graph in clinic, in order to determine if a prosthesis is performing better, worse or as expected. This would assist the learning curve in surgical technique and act as a guide to predict the potential need for future surgical intervention. Collection of a data set for a particular surgeon and implant would allow an accurate pre-printed graph to be developed for comparison (Fig. 17.3). The model could ultimately be used to compare measured wear rates between different surgeons or between units. Wear measurement can be quickly and easily performed in clinic with the use of a digital calliper, a light box and a transparent sheet containing concentric circles.

References

1. Brooker, A.F., Bowerman, J.W., Robinson, R.A., Riley Jr., L.H.: Ectopic ossification following total hip replacement. Incidence and a method of classification. J. Bone Joint Surg. Am. **55-A**, 1629–1632 (1973)
2. Gruen, T.A., McNeice, G.M., Amstutz, H.C.: "Modes of failure" of cemented stem-type femoral components: a radiographic analysis of loosening. Clin. Orthop. Relat. Res. **141**, 17–27 (1979)
3. DeLee, J.G., Charnley, J.: Radiological demarcation of cemented sockets in total hip replacement. Clin. Orthop. Relat. Res. **121**, 20–32 (1976)
4. Engh, C.A., Massin, P., Suthers, K.E.: Roentgenographic assessment of the biologic fixation of porous-surfaced femoral components. Clin. Orthop. Relat. Res. **257**, 107–128 (1990)
5. Nunn, D., Freeman, M.A., Hill, P.F., Evans, S.J.: The measurement of migration of the acetabular component of hip prostheses. J. Bone Joint Surg. Br. **71-B**, 629–631 (1989)
6. Livermore, J., Ilstrup, D., Morrey, B.: Effect of femoral head size on wear of the polyethylene acetabular component. J. Bone Joint Surg. Am. **72-A**, 518–528 (1990)
7. Ebramzadeh, E., Sangiorgio, S.N., Lattuada, F., Kang, J.S., Chiesa, R., McKellop, H.A., Dorr, L.D.: Accuracy of measurement of polyethylene wear with use of radiographs of total hip replacements. J. Bone Joint Surg. Am. **85-A**, 2378–2384 (2003)
8. Dowd, J.E., Sychterz, C.J., Young, A.M., Engh, C.A.: Characterization of long-term femoral-head-penetration rates: association with and prediction of osteolysis. J. Bone Joint Surg. Am. **82-A**, 1102–1107 (2000)

9. Martell, J.M., Berkson, E., Berger, R., Jacobs, J.: Comparison of two and three-dimensional computerized polyethylene wear analysis after total hip arthroplasty. J. Bone Joint Surg. Am. **85-A**, 1111–1117 (2003)

10. Hui, A.J., McCalden, R.W., Martell, J.M., MacDonald, S.J., Bourne, R.B., Rorabeck, C.H.: Validation of two and three-dimensional radiographic techniques for measuring polyethylene wear after total hip arthroplasty. J. Bone Joint Surg. Am. **85-A**, 505–511 (2003)

11. Sychterz, C.J., Yang, A.M., McAuley, J.P., Engh, C.A.: Two-dimensional versus three-dimensional radiographic measurements of polyethylene wear. Clin. Orthop. Relat. Res. **365**, 117–123 (1999)

Comparative In Vivo Wear Measurement of Conventional and Modern Bearing Surfaces in Total Hip Replacements by the Use of POLYWARE® Computerized System

18

Georgios Karydakis and Theofilos Karachalios

18.1
Introduction

Total hip arthroplasty operations represent an achievement of modern orthopaedics having given solution in millions of people with end stage degenerating disease of hip joint [34, 2]. The purpose of these operations is restoration of normal and painless function of the joint and maintenance of this outcome for many years.

Wear debris of bearing surfaces contributes to osteolysis leading to limitation of the longevity of total hip arthroplasties. Many alternative combinations of bearing surfaces have been developed in order to reduce the amount of wear produced.

A hard on hard bearing couple is an attractive solution due to minimum wear produced, but this may not be suitable for all patients. Ceramic on ceramic combinations have been noted to squeak [1] and require optimal placement in order to avoid the risk of neck to socket impingement [3]. Metal on metal combinations have been associated with pseudo-tumor formation [4] and increased metal ion release which may lead to DNA changes [5].

Hard on soft bearing couples apply to the majority of population and surgeons, but are associated with wear and osteolysis [6–9]. Metallic femoral head on polyethylene, introduced by Sir John Charnley, is the "gold standard", however, ceramic head on polyethylene

G. Karydakis (✉)
Orthopaedic Department,
251 Hellenic Airforce General Hospital,
3 Kanellopoulou Street, 15125
Athens, Hellenic Republic
e-mail: gkarid@med.uth.gr

T. Karachalios
Orthopaedic Department, Faculty of Medicine, School of Health Sciences,
University of Thessalia, Larissa, Hellenic Republic

K. Knahr (ed.), *Tribology in Total Hip Arthroplasty*,
DOI: 10.1007/978-3-642-19429-0_18, © 2011 EFORT

(introduced early in 1970s) demonstrates up to 20 times less wear than the former in hip simulators [10, 11] although it has a relatively low but considerable risk of fracture. On the other side, polyethylene underwent many improvements from ultra high molecular weight (UHMWPE) to cross-linked (XLPE) and extensive or highly cross-linked (HXLPE) poly-ethylene. These improved polyethylenes (particularly the latter) have shown reduced wear in combination with metallic heads in vitro [10, 12, 13] and in vivo [14–16].

It is possible that the combination of ceramic head with HXLPE or even better metallic head with the benefits of ceramic surface (such as Oxinium™, described below) with HXLPE should produce even lower amount of wear.

18.2
Aim

The aim of our study is to compare in vivo the two- and three-dimensional wear as well as the wear rate per year of three bearing surfaces in total hip replacements: ceramic head with conventional polyethylene (Ultra high molecular weight polyethylene, UHMWPE), ceramic head with highly cross-linked polyethylene (HXLPE) and Oxinium™ head with HXLPE.

18.3
Patients and Methods

Early in 2003 a prospective randomized study was introduced in our department to evalu-ate the in vivo wear performance of Oxinium™ on HXLPE. Two hundred and twenty one patients (244 hips), who underwent THA for osteoarthrosis of the hip, were randomised into four different groups. In group A (48 THAs) the bearing coupling used was ceramic on conventional PE, in group B (51 THAs) ceramic on highly cross-linked PE, in group C (55 THAs) Oxinium™ (28 mm) on highly cross-linked PE, in group D (60 THAs) Oxinium™ (32 mm) on highly cross-linked PE and in group E ceramic on ceramic (30 THAs). There were 144 female and 77 male patients. All operations were performed by a single surgeon. We used Synergy as femoral stem (Smith and Nephew, Memphis, USA), implanted with non-cementing technique. The acetabular component was the Reflection, a porous coated hemispherical shell (Smith and Nephew, Memphis, USA) implanted also with non-cementing technique in all cases. Patients in the five different groups were matched for age, sex, side and BMI. The follow-up observation period is 5.5 years on average (4–7 years). Two non-implant related failures were recorded in our series.

18.3.1
Polyethylene

We utilized two types of polyethylene: UHMWPE and HXLPE. Conventional polyethylene was implanted in 48 cases (34 females and 14 males) and highly cross-linked polyethylene in 166 cases [55 in combination with Oxinium head of 28 mm (39 females and 16 males), 60 in combination with Oxinium™ head of 32 mm (38 females and 22 males) and 51 with ceramic head (33 females and 18 males)].

18.3.2
Ceramic Head

In 99 cases (67 females and 32 males), we utilized ceramic heads (Biolox® forte, Ceramtec) of 28 mm. In 48 of them, ceramic head was combined with UHMWPE and in 51 with HXLPE. In 30 cases ceramic head was combined with ceramic insert.

18.3.3
Oxinium™ Head

Oxinium™ is an alloy consisted of Zirconium (97.5%) and Niobium (2.5%) that is subjected to heat in oxygen environment so that its surface is converted to a ceramic material, named oxidized zirconium (Zr_2O). This material is not a coating. It is expected to have the benefits of metallic head (fracture toughness) with the advantage of better polished surface that can be achieved with a ceramic surface.

In our study we included 115 cases with Oxinium™ heads (Smith and Nephew, Memphis, USA). Fifty-five of them had diameter of 28 mm (39 females and 16 males) and the remaining 60 cases had head diameter of 32 mm (38 females and 22 males). All Oxinium™ heads were combined with HXLPE.

The above information is summarized in Table 18.1.

Table 18.1 Combinations of bearing surfaces utilized in our study

Insert \ Head	Ceramic	Oxinium
UHMWPE	48	-
XLPE	51	28mm 51 / 60 / 32mm
Ceramic	30	-

18

18.3.4
In Vivo Wear Measurement

All patients are followed on regular intervals in a prospective manner. Both clinical (objective and subjective) and radiological data (standard anteroposterior and true lateral pelvis radiographs) is available for all patients (preoperatively, and at 3 weeks, 6 weeks, 3 months, 6 months, 12 months and every year thereafter). Every radiograph is digitized with a scanner. First the user creates a record for the patient with data concerning name, age, and dimensions of the implants. Then the radiograph is imported and analyzed by the program automatically, detecting the circles that correspond with the femoral head and acetabular shell and thus their centers (dual circle technique) (Fig. 18.1). The single point where the observer intervenes is to accept or to reject the analysis based on the contact or not of the circles to the periphery of the acetabular shell and femoral head (Fig. 18.2). After accepting the analysis, software extracts the results in a text file containing the linear wear in millimeters (mm), the volumetric wear in mm^3 and the volumetric wear rate in mm^3/year. These measurements are utilized to compare the three parameters mentioned before for every patient over time and thus for every combination of bearing surfaces.

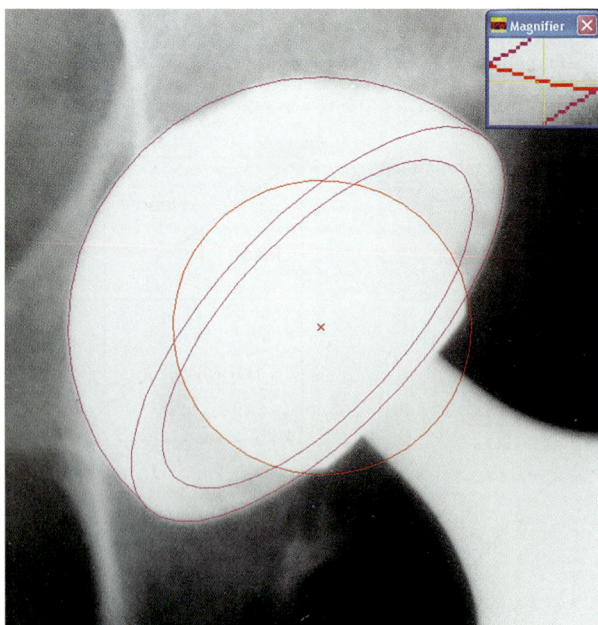

Fig. 18.1 Correct analysis of a radiograph: the model fit is satisfactory. The magnifier window ensures that the red circle is adjacent to the periphery of femoral head

Fig. 18.2 Incorrect analysis of a radiograph

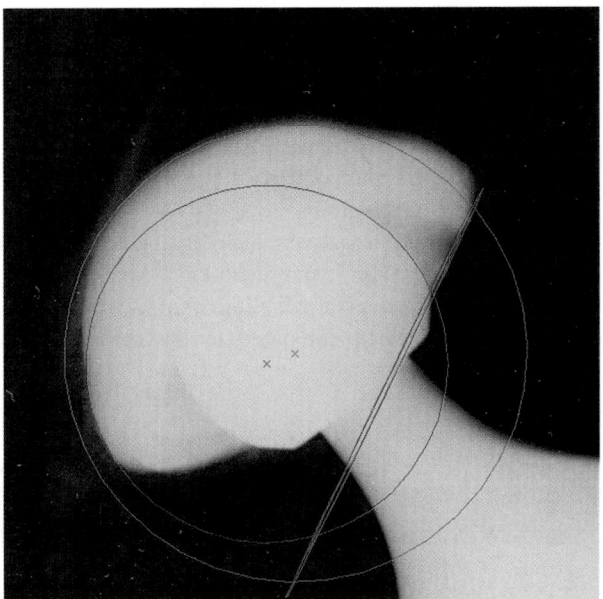

18.4
Statistics

Statistical analysis included the Wilcoxon signed rank test for non-parametric data and a two tailed *t* test to determine significant differences between the three groups. *P*-values at the level of 0.5 were considered as significant.

18.5
Results

The linear (two dimensional) wear for the combination of ceramic head on UHMWPE was found to be 1.051 mm during the first year, 1.07 mm during the first 2 years, 1.201 for 3 years, 1.428 for 4 years, 1.478 for 5 years and 1.508 for 6 years of follow-up. For the combination of ceramic head on HXLPE, the linear wear was 0.985, 1.2095, 1.896, 2.306, 2.356 and 2.425 respectively, whereas for Oxinium head on HXLPE, the linear wear was 0.52, 0.733, 0.618, 0.645, 0.635 and 0.648 (for 32 mm head) and 0.501, 0.711, 0.613, 0.663, 0.651 and 0.673 (for 28 mm head) respectively. All these differences were statistically significant ($p<0.01$).

18

The volumetric wear (mm³) and volumetric wear rate (mm³/year) for the three bearing couples had similar differences. All these data are summarized in Tables 18.2–18.4 and graphs in Figs. 18.3–18.5 depict the linear wear, volumetric wear and volumetric wear rate for the three bearing surfaces comparatively.

All three parameters (linear wear, volumetric wear and volumetric wear rate) were favorable for the Oxinium on HXLPE group followed by the ceramic on UHMWPE and the ceramic on HXLPE group.

Volumetric wear rate was very high for all groups during first year and had a substantial reduction thereafter. This is showed in Fig. 18.6 and will be explained below.

Both Oxinium groups of couplings showed statistically significantly less wear after the second year when compared to the ceramic on UHMWPE and ceramic on HXLPE.

Table 18.2 2D wear (mm) for all bearing surfaces

	First year	Second year	Third year	Fourth year	Fifth year	Sixth year
Ceramic – UHMWPE	1.051	1.07	1.201	1.428	1.478	1.508
Ceramic – XLPE	0.985	1.2095	1.896	2.306	2.356	2.425
Oxinium – XLPE 28 mm	0.5	0.7	0.6	0.66	0.65	0.67
Oxinium – XLPE 32 mm	0.52	0.733	0.618	0.645	0.635	0.648

Table 18.3 3D wear (mm³)

	First year	Second year	Third year	Fourth year	Fifth year	Sixth year
Ceramic – UHMWPE	530.963	536.187	492.708	754.891	790.9	810.5
Ceramic – XLPE	420.50	723	1110.616	1375.35	1420.2	1500.9
Oxinium – XLPE 28 mm	240.2	310.7	265.9	288.11	320.22	324.34
Oxinium – XLPE 32 mm	266.345	341.946	285.33	298.21	300.2	305.8

Table 18.4 3D wear rate (mm³/year)

	First year	Second year	Third year	Fourth year	Fifth year	Sixth year
Ceramic – UHMWPE	1729.067	404.14	301.23	298.21	305.8	310.5
Ceramic – XLPE	1721.247	378.75	394.7	361.63	365.8	372.7
Oxinium – XLPE 28 mm	1900.26	243.839	192.92	203.5	204.8	205.4
Oxinium – XLPE 32 mm	1710.32	190.54	210.4	205.4	206.23	210.98

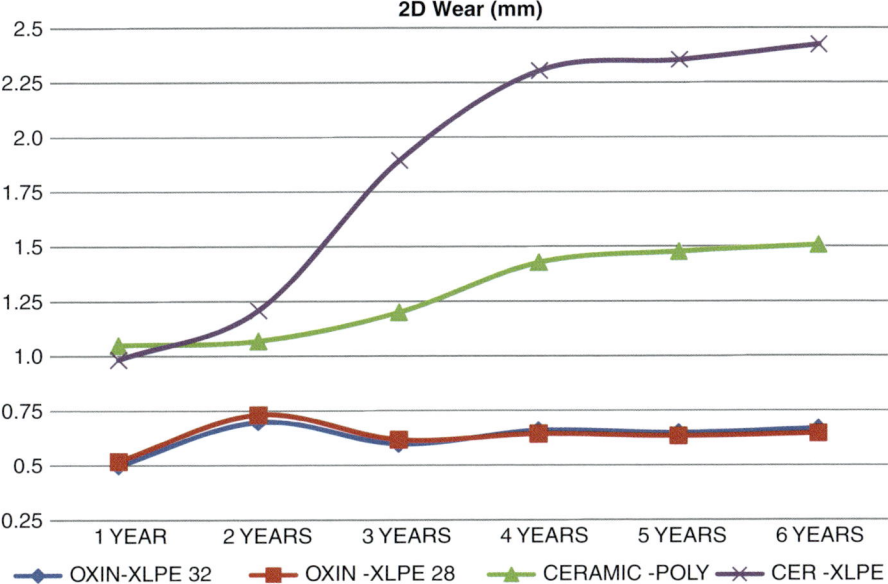

Fig. 18.3 2D (linear) wear in mm

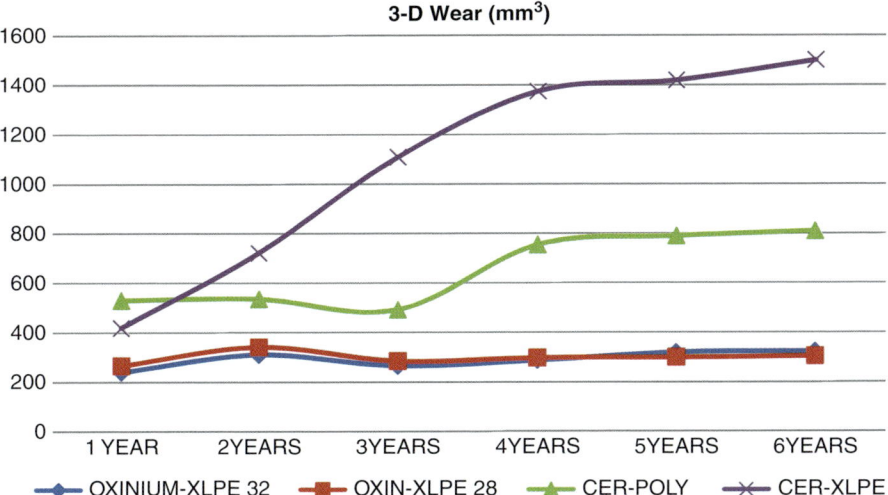

Fig. 18.4 3D (volumetric) wear in mm³

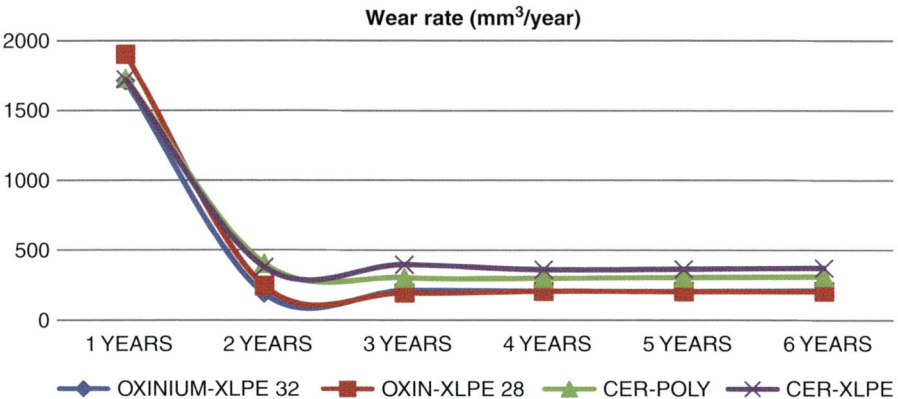

Fig. 18.5 Wear rate in mm³/year

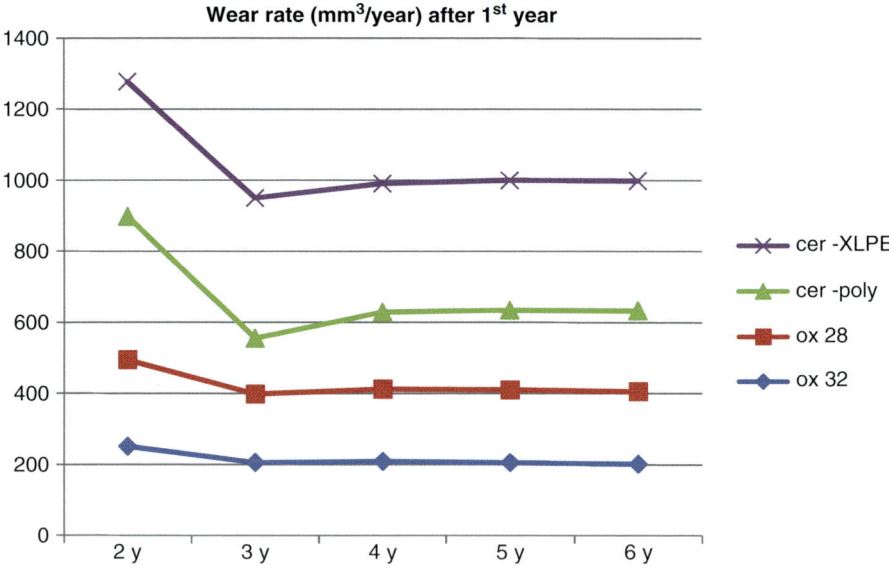

Fig. 18.6 Wear rate in mm³/year after first year

18.6
Discussion

The objective of this study was to compare in vivo three parameters of wear (linear, volumetric and volumetric rate) in four bearing couples in total hip arthroplasties: Ceramic on UHMWPE, ceramic on HXLPE and Oxinium of 28 and 32 mm on HXLPE. The results of our study demonstrate that the combination of Oxinium femoral head of both 28 and

Table 18.5 Accuracy and precision of various methods of wear analysis (From Martell [21])

	Accuracy (mm)	Precision (mm)
Livermore	0.18	1.75–2.18
EBRA	0.90	–
RSA	0.030–0.077	0.060
PolyWare	0.026–0.10	0.006–1.07

32 mm with highly cross-linked polyethylene as acetabular insert provides a significant lower amount of linear wear, volumetric wear and volumetric wear rate than the two other. The next combination of bearing coupling in terms of performance was found to be ceramic head on UHMWPE and the least efficient ceramic head on HXLPE.

In order to measure wear in vivo, many techniques have been developed: manual methods as Livermore had described [17, 18] and recently computer assisted techniques that are considered to have better accuracy [19, 20, 32]. There is a variety of the latter with different levels of accuracy and precision (Table 18.5) and with various disadvantages however. For example, RSA (Radiostereometric analysis), which is one of the most accurate technique today, has the disadvantage of the necessity of implanting indicators during surgery. Our choice was Polyware™ Auto 3D Digital, version 6.01 that was introduced by Devane et al. in 1995 [19, 22, 33] and developed by Draftware Developers Inc. The software is able to measure wear between hard on soft bearing surfaces, therefore we couldn't use it for group E (ceramic–ceramic bearing couple).This technique ensures better repeatability (× 10 times) than manual techniques such as Livermore's mentioned before [20].

Oxinium™ heads has some advantages in comparison to cobalt–chrome heads: it is a material with metal core and abrasion-resistant ceramic surface that is > 4 µm thick and has been shown to be continuous without pores or voids, chemically bonded to the underlying metal substrate and difficult to penetrate. It is associated with considerably less wear on hip simulators than comparable Co–Cr femoral heads articulating with conventional or cross-linked PE even when damaged [23], but there are no clinical data supporting it. It has also advantages over ceramic heads because of minimal risk of fracture. Although risk for ceramic femoral head fracture is relatively low, it is considerable when it occurs. Not only is reoperation necessary, but the outcome of the revision arthroplasty for ceramic femoral head fracture compromised because of retained ceramic fragments that have been shown to increase wear, to increase the development of osteolysis, and to increase the need for another revision surgery [24, 25].

During the first year we detected a high rate of wear for all bearing surfaces (Fig. 18.5). This is explained by creep of polyethylene, plastic deformation and initial bedding-in. Many other studies of in vivo wear measurement have remarked this phenomenon [14–16, 26–28]. Some of them use as time period for this 3 months [15], other 6 months [27] and most of them 12 months [10, 21, 25]. In order to accentuate the differences of wear rate between the bearing surfaces studied, in Fig. 18.6 we outline the wear rate after the first year.

18

The lowest amount of wear was produced by the combination of Oxinium™ with HXLPE. This finding is in accordance to the only published in vivo study [26] and to in vitro studies [29]. Additionally there was no statistically significant difference between the 28 and 32 mm Oxinium™ femoral heads and the study mentioned before [26] though it utilizes the two sizes of femoral heads (and one 36 mm), does not separate them, being unable to compare. This potentially allow for the use of bigger size of femoral head which approximates the normal kinematics improving range of motion [30] and reduces the risk for dislocation [31].

It would be expected that the use of HXLPE should produce lower amount of wear in comparison to UHMWPE whether combining the former with ceramic head or Oxinium™ head. Our results demonstrate that the combination of ceramic femoral head with HXLPE produce more wear than with UHMWPE. No reports have yet examined the clinical efficacy of the combination of highly cross-linked polyethylene cup with a ceramic head, except one recent study from Osaka, Japan that compares clinically the combination of HXLPE with Co–Cr or ceramic head, concluding that there is no difference between them [28]. Our result could be explained from the fact that HXLPE has characteristics that approximate hard surface. Another point to mention is that they derive from mid-term follow-up (mean 5.5 years) thus it is not safe to elicit definitive conclusions.

18.7
Conclusions

In conclusion, all bearing surfaces showed a high wear rate of a varying degree up to the second postoperative year. This reflects the initial bending in (plastic deformation – creep) of the head within the PE liner. After this, Oxinium™ on cross-linked PE showed the lowest wear rate which reflects a satisfactory in vivo behavior. No differences were observed when the 28 and 32 mm Oxinium™ heads were compared. The ceramic on conventional PE coupling showed satisfactory and comparable to historical controls wear rates. Surprisingly, the ceramic on this specific HXLPE showed the worst wear rate suggesting that all PEs and all ceramic on cross-linked PEs are not the same. This study reflects mid-term in vivo wear rates of these different couplings and long-term data is needed until definite conclusions can be made. Additionally, no cost-effectiveness data supporting the use of Oxinium™ on cross-linked PEs bearing coupling exist.

References

1. Mai, K., Verioti, M.K.C., Ezzet, K.A., Copp, S.N., Walker, R.H., Colwell Jr., C.W.: Incidence of 'squeaking' after ceramic-on-ceramic total hip arthroplasty. Clin. Orthop. Relat. Res. **468**, 413–417 (2010)
2. Ritter, M.A., Albohm, M.J., Keating, E.M., Faris, P.M., Meding, J.B.: Comparative outcomes of total joint arthroplasty. J. Arthroplasty **10**, 737–741 (1995)

3. Walter, A.: On the material and the tribology of alumina-alumina couplings for hip joint prostheses. Clin. Orthop. Relat. Res. **282**, 31–46 (1992)
4. Kwon, Y.M., Ostlere, S.J., McLardy-Smith, P., Athanasou, N.A., Gill, H.S., Murray, D.W.: Asymptomatic pseudotumors after metal-on-metal hip resurfacing arthroplasty prevalence and metal ion study. J. Arthroplasty (2010) [Epub ahead of print]
5. Ladon, D., Doherty, A., Newson, R., Turner, J., Bhamra, M., Case, C.P.: Changes in metal levels and chromosome aberrations in the peripheral blood of patients after metal-on-metal hip arthroplasty. J. Arthroplasty **19**(8 Suppl 3), 78–83 (2004)
6. Jasty, M.J., Floyd, W.E., Schiller, A.L., Goldring, S.R., Harris, W.H.: Localized osteolysis in stable, non-septic total hip replacement. J. Bone Joint Surg. Am. **68**, 912–919 (1986)
7. Willert, H., Bertram, H., Buchhorn, G.: Osteolysis in alloarthroplasty of the hip: the role of bone cement fragmentation. Clin. Orthop. Relat. Res. **258**, 108–121 (1990)
8. Willert, H., Bertram, H., Buchhorn, G.: Osteolysis in alloarthroplasty of the hip: the role of ultra high molecular weight polyethylene wear particles. Clin. Orthop. Relat. Res. **258**, 95–107 (1990)
9. Willert, H.G., Semlitsch, M.: Tissue reactions to plastic and metallic wear products of joint endoprostheses. Clin. Orthop. Relat. Res. **333**, 4–14 (1996)
10. Santavirta, S., Boehler, M., Harris, W.H., Konttinen, Y.T., Lappalainen, R., Muratoglu, O., Rieker, C., Salzer, M.: Alternative materials to improve total hip replacement tribology. Acta Orthop. Scand. **74**, 380–388 (2003)
11. Urban, J.A., Garvin, K.L., Boese, C.K., Bryson, K., Pedersen, D.R., Callaghan, J.J., Miller, R.K.: Ceramic-on-polyethylene bearing surfaces in total hip arthroplasty: seventeen to twenty-one-year results. J. Bone Joint Surg. Am. **83**, 1688–1694 (2001)
12. Muratoglu, O.K., Greenbaum, E.S., Bragdon, C.R., Jasty, M., Freiberg, A.A., Harris, W.H.: Surface analysis of early retrieved acetabular polyethylene liners: a comparison of conventional and highly crosslinked polyethylenes. J. Arthroplasty **19**, 68–77 (2004)
13. Muratoglu, O.K., Merrill, E.W., Bragdon, C.R., O'Connor, D., Hoeffel, D., Burroughs, B., Jasty, M., Harris, W.H.: Effect of radiation, heat, and aging on in vitro wear resistance of polyethylene. Clin. Orthop. Relat. Res. **417**, 253–262 (2003)
14. Beksac, B., Salas, A., Della Valle, A.G., Salvati, E.: Wear is reduced in THA performed with highly cross-linked polyethylene. Clin. Orthop. Relat. Res. **467**, 1765–1772 (2009)
15. Glyn-Jones, S., Isaac, S., Hauptfleisch, J., McLardy-Smith, P., Murray, D.W., Gill, H.S.: Does highly cross-linked polyethylene wear less than conventional polyethylene in total hip arthroplasty? A double blind, randomized, and controlled trial using roentgen stereophotogrammetric analysis. J. Arthroplasty **23**, 337–343 (2008)
16. Manning, D.W., Chiang, P.P., Martell, J.M., Galante, J.O., Harris, W.H.: In vivo comparative wear study of traditional and highly crosslinked polyethylene in total hip arthroplasty. J. Arthroplasty **20**, 880–886 (2005)
17. Karachalios, T., Hartofilakidis, G., Zacharakis, N., Tsekoura, M.: A 12- to 18-year radiographic follow-up study of Charnley low-friction arthroplasty. Clin. Orthop. Relat. Res. **296**, 140–147 (1993)
18. Livermore, J., Ilstrup, D., Morrey, B.: Effect of femoral head size on wear of the polyethylene acetabular component. J. Bone Joint Surg. Am. **72A**, 518–528 (1990)
19. Ebramzadeh, E., Sangiorgio, S.N., Lattuada, F., Kang, J.S., Chiesa, R., McKellop, H.A., Dorr, L.D.: Accuracy of measurement of polyethylene wear with use of radiographs of total hip replacements. J. Bone Joint Surg. Am. **85A**, 2378–2384 (2003)
20. Martell, J.M., Berdia, S.: Determination of polyethylene wear in total hip replacements with use of digital radiographs. J. Bone Joint Surg. Am. **79A**, 1635–1641 (1997)
21. Martell, J.: Clinical wear assessment. In: Callaghan, J., Rosenberg, A., Rubash, H. (eds.) The Adult Hip, 2nd edn. Lippincott Williams & Wilkins, Philadelphia (2007)
22. Devane, P.A., Horne, J.G.: Assessment of polyethylene wear in total hip replacement. Clin. Orthop. Relat. Res. **369**, 59–72 (1999)

23. Benezra, V., Mangin, S., Treska, M., et al.: Microstructural investigation of the oxide scale on Zr-2.5 Nb and its interface with the alloy substrate. Mater. Res. Soc. Symp. Proc. **550**, 337–342 (1999)
24. Allain, J., Roudot-Thorval, F., Delecrin, J., Anract, P., Migaud, H., Goutallier, D.: Revision total hip arthroplasty performed after fracture of a ceramic femoral head: a multicenter survivorship study. J. Bone Joint Surg. **85A**, 825–830 (2003)
25. Hasegawa, M., Sudo, A., Uchida, A.: Cobalt-chromium head wear following revision hip arthroplasty performed after ceramic fracture – a case report. Acta Orthop. **77**, 833–835 (2006)
26. Garvin, K.L., Hartman, C.W., Mangla, J., Murdoch, N., Martell, J.M.: Wear analysis in THA utilizing oxidized zirconium and crosslinked polyethylene. Clin. Orthop. Relat. Res. **467**(1), 141–145 (2009)
27. Mutimer, J., Devane, P.A., Adams, K., Horne, J.G.: Highly crosslinked polyethylene reduces wear in total hip arthroplasty at 5 years. Clin. Orthop. Relat. Res. **468**, 3228–3233 (2010)
28. Nakahara, I., Nakamura, N., Nishii, T., Miki, H., Sakai, T., Sugano, N.: Minimum five-year follow-up wear measurement of Longevity highly cross-linked polyethylene cup against cobalt-chromium or zirconia heads. J. Arthroplasty **8**, 1182–1187 (2010)
29. Good, V., Ries, M., Barrack, R.L., Widding, K., Hunter, G., Heuer, D.: Reduced wear with oxidized zirconium femoral heads. J. Bone Joint Surg. Am. **85**(Suppl 4), 105–110 (2003)
30. Burroughs, B.R., Hallstrom, B., Golladay, G.J., Hoeffel, D., Harris, W.H.: Range of motion and stability in total hip arthroplasty with 28-, 32-, 38-, and 44-mm femoral head sizes: an in vitro study. J. Arthroplasty **20**(1), 11–19 (2005)
31. Amlie, E., Høvik, Ø., Reikerås, O.: Dislocation after total hip arthroplasty with 28 and 32-mm femoral head. J. Orthop. Traumatol. **11**(2), 111–115 (2010)
32. Borlin, N., Thien, T., Karrholm, J.: The precision of radiostereometric measurements. Manual vs. digital measurements. J. Biomech. **35**, 69–79 (2002)
33. Devane, P.A., Bourne, R.B., Rorabeck, C.H., Hardie, R.M., Horne, J.G.: Measurement of polyethylene wear in metal-backed acetabular cups. I. Three-dimensional technique. Clin. Orthop. Relat. Res. **319**, 303–316 (1995)
34. Rissanen, P., Aro, S., Slätis, P., Sintonen, H., Paavolainen, P.: Health and quality of life before and after hip or knee arthroplasty. J. Arthroplasty **10**, 169–175 (1995)

Index

K. Knahr (ed.), *Tribology in Total Hip Arthroplasty*,
DOI: 10.1007/978-3-642-19429-0, © 2011 EFORT

Printing: Ten Brink, Meppel, The Netherlands
Binding: Stürtz, Würzburg, Germany